Principles of Emergency Management and Emergency Operations Centers (EOC)

Principles of Emergency Management and Emergency Operations Centers (EOC)

Michael J. Fagel, PhD, CEM

CRC Press
Taylor & Francis Group
Boca Raton   London   New York

CRC Press is an imprint of the
Taylor & Francis Group, an **informa** business

Figure 4.2 is reprinted with Permission from the NFPA 704-2007, System for the Identification of the Hazards of Materials for Emergency Response, Copyright © 2006, National Fire Protection Association, Quincy, MA. This reprinted material is not the complete and official position of the NFPA on the referenced subject, which is represented only by the standard in its entirety.

CRC Press
Taylor & Francis Group
6000 Broken Sound Parkway NW, Suite 300
Boca Raton, FL 33487-2742

© 2011 by Taylor and Francis Group, LLC
CRC Press is an imprint of Taylor & Francis Group, an Informa business

No claim to original U.S. Government works

Printed in the United States of America on acid-free paper
10 9 8 7 6 5 4 3 2 1

International Standard Book Number: 978-1-4398-3851-8 (Hardback)

Library of Congress Cataloging-in-Publication Data

Fagel, Michael J.
    Principles of emergency management and emergency operations centers (EOC) / [edited by] Michael J. Fagel.
        p. cm.
    Includes bibliographical references and index.
    ISBN 978-1-4398-3851-8 (hardcover : alk. paper)
    1. Emergency management. I. Title.

HD49.F34 2011
658.4'77--dc22   2010043390

Visit the Taylor & Francis Web site at
http://www.taylorandfrancis.com

and the CRC Press Web site at
http://www.crcpress.com

It's been ten years since those devastating attacks in September 2001 on the World Trade Center and the Pentagon, and the horrific crash in Shanksville, Pennsylvania. How far have we come in this last decade?

This book is dedicated to my family, my parents, and my children. As my career expanded in these emergency management fields, my public safety, my response, has come at a cost to my family. To my wife, Patricia, who has supported me in my emergency management ventures worldwide, I must express a sincere debt of gratitude to her and to all of my family.

It is my sincere hope that as you read this text, please remember the people, the families, the rescuers, who are impacted daily by events that are manmade and naturally occurring. Terrorism has many shapes and many faces. It is my desire that what you will glean from inside this textbook will help make you better prepeared to cope. Take care of yourself, your family, and the public whom we serve.

I say simply thank you.

Contents

Foreword

Principles of Emergency Management and Emergency Operations Centers (EOC) is an essential tool that should be in every new emergency manager's tool belt and on every seasoned emergency manager's reference shelf. Mike Fagel has updated and enhanced the text from its first edition, bringing it current with the rapidly evolving homeland security and emergency management professions. Since the establishment of the Federal Emergency Management Agency (FEMA) in 1979, emergency management has continued to rapidly evolve to embrace the many changes in preparedness, response, and recovery that have occurred in order to prepare the nation to respond to any threat or malady. Along the way, the profession had to learn to deal with acts of terrorism in addition to the usual blend of natural and human-caused incidents. In 2002, the profession adopted what can arguably be the most significant change in emergency management with the creation of the U.S. Department of Homeland Security. In the years since, states and local jurisdictions have also established homeland security-related agencies or added additional responsibilities to already busy emergency management departments.

In addition to the many changes to emergency management caused by an evolving profession, emergency management professionals have also had to deal with several high consequence events within the nation, not to mention those of our friends and neighbors across the globe to which we responded. The recommendations of the 9/11 Commission and the Post-Katrina Emergency Reform Act both were instigators of further change to the profession as a result of looking back on how the nation responded to these two high-consequence events. Certainly one of the key findings in many after action reviews (AARs) following significant events is the need to ensure effective command, control, and coordination across a myriad of response and recovery agencies at all levels. Over the past many years, the emergency operations center (EOC) has become the symbol of effective and efficient emergency management. The EOC is a fairly simple concept but often proves to be anything but simple to operate, especially over long activation periods. Having a thorough understanding of how an EOC should operate within the guidance of various federal and national programs (such as the National Incident Management System [NIMS], the National Response Framework [NRF], and others) is the foundation for sound EOC management. But, to achieve excellence, the emergency manager must go beyond the basics and mandates. She or he must seek best practices and must study lessons from fellow emergency managers who have had to activate their EOCs in times of crisis. Emergency managers have learned that not all EOC activations will occur as the result of a "big bang" type of event, but they can have value in terms of other types of incidents such as public health incidents (e.g., pandemics). Fortunately, most jurisdictions do not have to face national level high consequence events on a regular basis. When they do, emergency managers have learned to assess "how things went" through detailed AARs. The absence of regular activations of the EOC does relieve the emergency manager of the obligation to regularly evaluate his or her operations and how they contribute to the effective response to a crisis. Consequently, exercises that are properly planned, conducted, and evaluated have become an essential tool of today's emergency manager to ensure that the EOC and emergency management operations are always ready for activation.

Mike Fagel's updated text should be in every emergency manager's bookshelf and should develop the look and appearance of a go-to resource. It should become a little worn, perhaps a bit frayed; and it should begin to sport highlighter markups and page turndowns—all signs of the good tool put to use by great professionals.

Rick C. Mathews, MS
Director
National Center for Security & Preparedness
Rockefeller College of Public Affairs and Policy
University at Albany, State University of New York

Preface

When I started in emergency management, it was called civil defense. Today we call it homeland security. Regardless of what the process is called in your jurisdiction or organization, at its foundation there is an "all risks–all hazards" approach to planning, preparedness, mitigation, and response.

In this text, I have collaborated with some of my most trusted colleagues in this field to share these following pages with you.

My own experiences in over 40 years of emergency management and public safety (emergency services sector) have helped shape and synthesize my focus, learning from the past and helping to create the future emergency managers and planners of the twenty-first century. I have (and continue to do so) contributed to and participated in many local, state, national, and international tabletop and full-scale exercises and symposiums to prepare jurisdictions for a more ready response to any emergency.

I started my years in fire, rescue, law enforcement, emergency medical services, and civil defense (emergency mangement) with Cadillac ambulances and Rambler patrol cars with walkie-talkies that weighed five pounds. I then had deployments to the Oklahoma City Bombing in 1995 and the World Trade Center attack in 2001, as well as numerous Federal Emergency Management Agency (FEMA) deployments to floods, hurricanes, ice storms, and tornadoes. I have traveled from Navy bases in the Pacific to Army bases in Alaska for the U.S. Army Soldier Biological Chemical Command at Aberdeen Proving Grounds. I was also a part of the FBI National Domestic Preparedness Office four years before the Department of Homeland Security was created. I have also trained others in the distant lands of the Middle East developing FEMA-like organizations and emergency operations centers.

When I started teaching in 1976 at Waubonsee Community College, little did I realize where my future would take me. I am also currently at the Argonne National Laboratory, Infrastructure Assurance Center working in critical infrastructure/key resources, as well as teaching at Northwestern University, Louisiana State University, Eastern Kentucky University, and Benedictine University.

All of these events have shaped my life, my career, and my passion. Last, but certainly not least, without the support of my family, I would not be the field practitioner, the teacher, or the person that I am today. This book is for all of you.

Oklahoma City Bombing, April 19, 1995
World Trade Center Attacks, September 11, 2001
Hurricane Katrina, August 29, 2005
We can never forget!

For the future, let us all learn and make our world a safer place.

Acknowledgments

My professional experiences have been enhanced by the many people that I have had the distinct honor and opportunity to work with over my many decades in the emergency services sector. This text is a collaboration among some of those people and would not be in your hands today without their assistance. To all of them and those that I may have not mentioned here, but have learned from over the years, I thank you.

The contributors to this book are my trusted colleagues and are all dedicated professionals who have partnered with me to help create an authoritative text for those engaged in emergency planning, emergency operations, preparedness, response, and mitigation. A sincere debt of gratitude is owed to them all for their advice, counsel, and friendship.

Author

Michael J. Fagel, Ph.D.,CEM, has more than 40 years of experience in emergency management, fire and emergency medical service, public health, law enforcement, bio-terrorism awareness and prevention, as well as corporate safety, security and threat risk management.

He served as an officer on the North Aurora Fire Department for over 28 years. He was a FEMA Reservist and responded to many federally declared emergencies, including deployment to the Oklahoma City Bombing in 1995 and the World Trade Center Attacks in 2001. He has provided WMD instruction to federal facilities for the U.S. Army, as well as providing technical support to state and county agencies on the CDC's SNS Planning. He has served as a subject matter expert to the National Guard Bureau and provided national EOC planning and CONOPS for the USA and Middle East venues, including the Salt Lake City Olympics.

He teaches at Northwestern University-Chicago, Benedictine University, Eastern Kentucky University, Louisiana State University's National Center for Biomedical Research and Training (NCBRT) Response to Agricultural Terrorism Training Program, as well as being an SME for the National Center for Security and Preparedness (NCSP) at the University at Albany, State University of New York. Fagel was a technical advisor and taught at the University of Chicago Masters in Threat Response Program. Fagel is also currently a Homeland Security Analyst at the Argonne National Laboratory engaged in the protection of Critical Infrastructure.

He is also a member of the Northern Illinois Critical Incident Stress Debriefing team, the International Association of Fire Chiefs Committee on Safety and Health, and was a member of their Terrorism Committee. He has published hundreds of articles and contributed to or published multiple textbooks on Emergency Management and safety topics.

He can be reached at mjfagel@aol.com.

Contributors

Lucien G. Canton, CEM, is an independent management consultant specializing in strategic planning for crisis. Prior to starting his own company, Canton served as the director of emergency services for the city and county of San Francisco and as an emergency management programs specialist and chief of the Hazard Mitigation Branch for FEMA Region IX. A popular speaker and lecturer, he is the author of the best-selling *Emergency Management: Concepts and Strategies for Effective Programs* used as a textbook in many higher-education courses. He can be contacted at LCanton@LucienCanton.com.

Robert J. Coullahan, CEM, CPP, CBCP, CHS V, is president and chief operating officer of Readiness Resource Group Incorporated in Las Vegas, Nevada. Coullahan has more than thirty-five years of professional experience in emergency management and critical infrastructure protection for government and commercial organizations. He is board certified in emergency management, security management, business continuity management, and homeland security. He founded Readiness Resource Group (www.readinessresource.net), which supports national preparedness and public safety programs. He served for twenty years with Science Applications International Corporation including assignment as senior vice president. He served five years as vice president for government and international programs with a university consortium that was named World Data Center for Human Interactions in the Environment, based upon U.S. National Research Council peer review. He serves on ANSI and ASIS Business Continuity Management Standards Committees. He holds the master of science in telecommunications and the master of arts in security management–forensic sciences, and is a graduate of the University of California and George Washington University. He can be contacted at coullahan@readinessresource.net.

Randall C. Duncan, MPA, CEM, has been a local government emergency manager since 1986, working at both the county and municipal levels of government in Kansas and Oklahoma. Duncan has been involved in administering nearly a dozen presidential declarations of disaster ranging from floods and tornadoes to ice storms. In addition, he provided support to the New York City Fire Department from September 18 to 28, 2001, at the incident command post for the World Trade Center. He has also been involved in other projects ranging from the Biological Warfare–Improved Response Program (BWIRP) to serving as an evaluator for the City of Seattle Emergency Operations Center in Topoff II. He is a past president of the USA Council of the International Association of Emergency Managers (IAEM-USA) and currently serves as its Government Affairs Committee chair. Duncan also serves as adjunct faculty for the Emergency Management Institute in Emmitsburg, Maryland, and Park University in Parkville, Missouri. He can be contacted at rduncan21@cox.net.

Ron Fisher is the deputy director of the Infrastructure Assurance Center (IAC) at Argonne National Laboratory. His responsibilities include technical support in many areas of critical infrastructure assurance to the Department of Homeland Security (DHS), Department of Energy, and Department of Defense. Fisher served as a senior consultant to the National Petroleum Council on oil and natural gas infrastructure vulnerabilities and for the President's Commission on Critical Infrastructure Protection. He currently serves as the IAC coordinator for the DHS Office of Infrastructure Protection support activities, which includes conducting field assessments, vulnerability assessment methodology development, risk analysis, and alignment to the National Infrastructure Protection Plan. Mr. Fisher is the author of several interdependencies papers and studies. He can be contacted at refisher@anl.gov.

K. R. Juzwin, PsyD, is a clinical psychologist and associate professor of clinical psychology at Argosy University-Schaumburg. She specializes in first responder mental health, high-risk personnel assessment, critical incident response, disaster mental health, and trauma. Juzwin is the regional coordinator for mental health response for the Illinois Medical Emergency Response Team (IMERT), a responder on the Illinois Disaster Assistance Team (D-MAT IL-2), and is the current education coordinator and has also been a coordinator on the Northern Illinois Critical Incident Stress Management Team (NICISM). On the IMERT team, she has written the training curriculum for mental health responders, and trains personnel related to mental health issues in crisis and disaster response. Clinically Juzwin focuses on trauma, forensic trauma, self-injury, eating disorders, and high-risk patients. She is active in education, consultation, writing and research in these areas. She is the director of Self-Injury Recovery Services at Alexian Brothers Behavioral Health Hospital in Hoffman Estates, Illinois, which is the only JCAHO-DSC certified program for self-injury in the country. Juzwin is the chief psychologist for COPS and FIRE Testing Service, and conducts outcome research, assessment, protocol development, and training and supervision in the area of testing and assessment for high risk hiring for law enforcement, fire/EMS, and emergency dispatch personnel. At Argosy University, Dr. Juzwin teaches testing and assessment, ethics, professional development; as well as police psychology, trauma, forensic trauma, eating disorders, self-injury and suicide assessment courses. She can be contacted at kjuzwin@argosy.edu.

Stephen J. Krill, Jr., is a senior associate at Booz Allen Hamilton in McLean, Virginia. Krill has twenty-two years of professional experience in emergency management, covering mitigation, preparedness, response, and recovery, for government, industry, and nongovernment organizations. He responded to Y2K, 9/11, and Hurricanes Katrina and Rita, and served as lead planner for national-level exercises. Krill is pursuing a PhD in engineering management with a focus in crisis, disaster, and risk management. He is a member of emergency management standards development organizations, such as ASTM, ANSI, and ISO. Krill previously served as an adjunct instructor at Northwestern University for transportation terrorism preparedness and presented at emergency management conferences held in both the United States and Europe. He led the development of three Reports to Congress, national guidance on critical infrastructure protection, and published several peer-reviewed journal articles. Krill holds professional certifications in emergency management, business continuity, and project management. He can be contacted at krill_stephen@bah.com.

Rick C. Mathews, MS, is the director of the National Center for Security and Preparedness (NCSP), an affiliate of the Rockefeller College of Public Affairs and Policy at the University at Albany, State University of New York. Prior to his current position at the NCSP, Mathews served as the assistant director for research and development at the National Center for Biomedical Research and Training (NCBRT), Academy of Counter-Terrorist Education at Louisiana State University. On April 19, 1995, he participated in the rescue efforts following the explosion of the Alfred P. Murrah Federal Building in Oklahoma City. He served as a technical consultant to the Louisiana Department of Health's Emergency Preparedness Planner during the response and recovery phases of Hurricane Katrina, dealing particularly with issues around the establishment and operation of the field hospital. In September 2008, he served as a technical consultant to the planning sector of the Louisiana Governor's Office of Homeland Security and Emergency Preparedness following Hurricane Gustav. His professional experience includes more than thirty years in emergency medical services, hospital administration, emergency preparedness, counterterrorism, and homeland security. He is a member of the International Association of Emergency Managers, the International Medical and Rescue Association, Protect New York, the National Intelligence Education Foundation, InfraGard, and the SESHA Environmental Safety & Health Association of New York. He can be contacted at rmathews@uamail.albany.edu.

James Peerenboom is director of the Infrastructure Assurance Center at Argonne National Laboratory. He has been actively engaged in systems analysis, decision and risk analysis, and advanced modeling and simulation activities for more than twenty-five years. For the past fifteen years, he has focused on critical infrastructure protection and homeland security issues, providing technical support to the Departments of Homeland Security, Energy, and Defense. He is the author of numerous publications on infrastructure interdependencies and received a PhD from the University of Wisconsin-Madison. He also serves as the associate director of the Decision and Information Sciences Division at Argonne National Laboratory. He can be contacted at jpeerenboom@anl.gov.

Thomas D. Schneid is a tenured professor in the Department of Safety, Security and Emergency Management (formerly Loss Prevention and Safety) at Eastern Kentucky University and serves as the graduate program director for the online and on campus master of science degree in safety, security and emergency management. He has worked in the safety and human resource fields for over thirty years at various levels including corporate safety director and industrial relations director. Schneid has represented numerous corporations in Occupational Safety and Health Administration (OSHA) and labor-related litigations throughout the United States. He has earned a BS in education, an MS and CAS in safety, an MS in international business, and a PhD in environmental engineering as well as his JD from West Virginia University and LLM from the University of San Diego. He is a member of the bar for the U.S. Supreme Court, 6th Circuit Court of Appeals and a number of federal districts as well as the Kentucky and West Virginia Bar. He has authored or coauthored fifteen texts including *Corporate Safety Compliance, Americans With Disabilities Act Handbook, Legal Liabilities for Safety and Loss Prevention Professionals and Fire Law* as well as over 100 articles on safety, fire, EMS, law, and related topics. He can be contacted at Tom.Schneid@eku.edu.

Nicholas Staikos, AIA, principal of Staikos Associates Architects, has over thirty years experience in the design of technologically complex projects including mission critical facilities for emergency management serving government and industry. Under his leadership, Staikos Associates Architects has completed numerous mission critical facilities development projects for all levels of government. The firm has guided the development of EOC design projects beginning with assessment of needs, criteria evaluation, architectural programming, and design through to technology deployment and occupancy. Staikos Associates Architects' projects have been recognized for their operational effectiveness and have received regional and national design awards as well as international recognition. Additionally, the firm's work at the Montgomery County (Maryland) Emergency Communications Center and the State of Delaware's State Emergency Operations Center were among the first, if not the first, such centers to integrate emergency management, transportation management and public safety on a physical as well as functional level. The firm's market segments include public safety and transportation management for projects such as 911 call centers, fusion centers, and transportation management centers. Staikos has authored articles on integrating security into computer facilities design and site planning. He can be contacted at NStaikos@Staikos.com.

Michael Steinle is the senior project manager for Tetra Tech EMI. Steinle has over twenty years of experience in project management; security; emergency response; and environmental, health, and safety stewardship. He has led many projects to assist in mitigating biological, chemical, physical, and radiological hazards, both in the public and private sectors. He also served on a multidisciplinary team charged by the Department of Army with assessing training efficacy for civilians and soldiers being deployed to Afghanistan and Iraq. Steinle served as the project manager on several large-scale state bioterrorism and public health planning projects. Steinle served as the technical lead on a Solar Cities Initiative in Boston and New York. The project was initiated to provide resiliency in emergency and homeland security operations through use of solar energy. He has also led security and emergency preparedness initiatives at airports, marine ports, and other transportation infrastructure. He has served for three years on the Planning Board of the Federal Bureau of Investigation's International Symposium on Agroterrorism (ISA) and served on a team that developed a script for a pandemic influenza video shot in Vietnam for the 2008 ISA. Steinle also served as the chair of the ASIS International Agriculture and Food Security Council from 2006 to 2008 and is an active member of the board of directors for the Safety and Health Council of Western Missouri and Kansas. He can be reached at msteinle@gmail.com.

Reviewers

Charles R. Blaich served the FDNY for 30 years. As Deputy Chief of the New York City Fire Department, he served as the Logistics Chief of the World Trade Center Task Force from September, 2001 to February, 2002, responsible for supporting recovery, fire fighting, GIS mapping, mortuary activities, care and feeding of the emergency and civilian workers, as well as emergency medical support to the Ground Zero workers. He was responsible for all logistic planning and activities supporting the entire array of emergency and support organizations, including police, EMS, Federal & State. He is a recognized expert on the interaction and management of the various levels of government and agencies necessary to successfully respond to disasters and terrorist incidents. His articles on building construction and collapse as well as emergency operations have appeared in many professional journals. He has lectured on the topics of disaster operations in many conferences and seminars worldwide.

Upon retirement, he accepted the position with the Raytheon Company as the *Director, Preparedness and Response (2003-2009)* in the Raytheon Homeland Security Strategic Business Area (SBA) at its inception. He assumed the position of *Director, Business Development Integrated Security Systems* in 2009 with BAE Systems, Inc. to explore expanding BAE Systems capabilities into the Public Safety Area of Homeland Security. He is currently in engaged as a private consultant.

Colonel Blaich served in both the Active Duty and Reserve United States Marine Corps from 1968-1998. His service included tours of duty in Vietnam; command of two battalions including the 2nd Battalion, 25th Marines Infantry Battalion in support of Operation Desert Storm. Awards include: Meritorious Service Medal, Two Navy Commendation Medals with combat "V", Combat Action Ribbon, National Defense Service Medal, as well as, several campaign and unit citation decorations. He can be contacted at: crblaich@gmail.com.

Chad Bowers is vice president of Bold Planning Solutions, a technology company focused on continuity planning, based in Nashville, Tennessee. Bowers has been actively engaged in the emergency management industry since 2001 and is an expert in the field of continuity of operations planning (COOP) for federal, state, and local government organizations. Bowers has served as senior project manager for dozens of COOP projects including statewide COOP initiatives for the State of Vermont, State of South Dakota, and the California Superior Courts. Currently, Bowers serves as senior technical manager for the development and direction of Bold Planning Solutions' highly acclaimed Web-based COOP system, EMplans. com. He can be contacted at Chad@BoldPlanning.com.

David K. Brannegan is Deputy Director of Special Assignments and Experts, Infrastructure Assurance Center (IAC) at Argonne National Laboratory. He has considerable experience in the non-profit, for-profit, and government sectors. He currently provides a wide range of support to multiple Department of Homeland Security (DHS) component offices in the measurement and reporting of protection measures and the expansion of fusion center capabilities. In his previous position as a Branch Chief at DHS, he oversaw the design, development, and implementation of over 20 individual technical assistance programs that enhanced state and local preparedness capabilities nationwide. He can be contacted at dbrannegan@anl.gov.

J. Howard Murphy is a veteran emergency manager, emergency responder, and national security program manager and senior analyst. He has responded to numerous disasters and commanded the U.S. Army's first CBRNE Incident Response Force. He is a veteran emergency manager, emergency responder, and national security program manager and senior analyst. His major areas of concentration are national security, weapons of mass destruction counterproliferation, and complex emergency management operations. Murphy has served as a chief officer, emergency manager, and commissioner within civilian emergency services and emergency management agencies; commanded several high-priority Department of Defense specialized teams; and cofounded and served on several nongovernmental organizations focused on disaster response and recovery operations globally. He can be contacted at jhmurphy@charter.net.

Chapter 1

Introduction

Michael Fagel and Stephen Krill

Contents

Disasters happen. No one is immune to them or the devastation that they can bring to communities of any size. *Disaster* applies to a variety of events, each with varying magnitudes and of varying natures. The Federal Emergency Management Agency (FEMA) defines a disaster as

> an occurrence of a severity and magnitude that normally results in deaths, injuries, and property damage and that cannot be managed through the routine procedures and resources of government.

Because of the unpredictable nature of disasters, it is essential to plan for such events. Only through planning is it possible to effectively respond to and mitigate against the potential effects of disaster.

Types of Disasters

Disasters usually fall into one of four categories: natural, manmade, deliberate, or accidental. In the United States, the most frequently occurring disasters are natural. Because natural disasters are not caused by the actions of humans, they are the most difficult to prevent. However, because most, although not all, natural disasters (e.g., weather related) are predictable, they should be the easiest to plan for.

Because it is impossible to stop natural disasters from occurring, emergency planners and responders must increase their awareness of how and why they occur. At the same time, they must analyze the risk in their communities so that they can focus their efforts on the highest risk or highest impact hazards. For example, Florida residents need not prepare for paralyzing blizzards. They must be prepared for hurricanes, however. On the other hand, New England residents need to be prepared for both blizzards and hurricanes as well as a multitude of other hazards.

Manmade disasters fall into two categories: deliberate and accidental. To identify a single cause of any manmade disaster would be nearly impossible. For as many types of manmade disasters that exist, there are a hundred reasons why. Whereas natural disasters are caused by the forces of nature and can only be predicted, not prevented, manmade disasters occur through preventable, sometimes deliberate, and often malicious acts.

Not all manmade disasters are deliberate, however. Many can be categorized as accidental and completely unexpected. As we continue to enter the era of technology, an increasing number of technologically sophisticated incidents will continue to occur, and community disaster plans must take into account

Table 1.1   Types of Disasters

Natural	Manmade	Deliberate	Accidental
Hurricanes	Nuclear accidents	Terrorism	Highway accidents
Tornados	Hazmat incidents	Strike violence	Rail accidents
Earthquakes	Explosions	Sabotage	Aircraft accidents
Snow storms		Bombings	Industrial mishaps
Ice storms		Riots and civil disturbances	
Floods			

the potential for an increasing number of incidents involving toxic spills, radio-logical releases, and other accidents. The events could be highly disastrous in densely populated areas, and as communities continue to grow, these events could have consequences that expand beyond the realm of anything imaginable in today's society.

Table 1.1 represents the types of disasters that fall into the categories just presented. This list is not intended to be inclusive, and certainly there are other potential disasters that could strike any community. For this reason, when the time comes to create or revise the community's disaster plan, it is critical that planners be prepared to protect their communities from these and other potential disasters that may occur at any time, with or without warning.

Each of these situations poses its own threat to communities and challenges to emergency response personnel. Some are relatively easy to predict, whereas others may occur with virtually no warning. Some will devastate a community for weeks or months, whereas others may pack a powerful punch, then be over, leaving devastation in their wakes.

It can be expected that most disasters, regardless of type or cause, will bring about some degree of death, injury, and property damage. Depending on the circumstances, disruption in communications systems, power, and water supplies; contamination of air, food, and water; and a multitude of other problems may also occur.

Phases of Disaster

Each type of disaster and its subsequent response effort—regardless of how, when, or why it takes place—can be expected to consist of a number of phases that are fairly consistent. The success of a community's response plan depends greatly upon the planner's understanding of these phases and how they are incorporated into the overall disaster plan.

The initial phase of a disaster is the *warning* phase. It is during this phase that emergency officials have the best opportunity to provide the public with disaster-related information. It is also during this phase that an evacuation or shelter in place, if necessary, will be initiated.

The second phase of a disaster is the *threat and impact* phase. This phase is typically followed by the *inventory, rescue, remedy,* and *restoration* phases.

Levels of Severity

Within each type of disaster, varying levels of severity will exist, each with its own response requirements. There are three levels of disaster, and even the lowest level can require the involvement of state and federal officials.

Level I Disaster—A localized multiple-casualty incident wherein local medical resources are available and adequate to provide for field medical treatment and stabilization, including triage. The patients will be transported to the appropriate medical facility for further diagnosis and treatment.

Level II Disaster—A multiple-casualty emergency where the large number of casualties and/or lack of medical care facilities are such as to require multi-jurisdictional medical mutual aid.

Level III Disaster—A mass-casualty emergency wherein local and regional medical resources are exceeded. Deficiencies in medical supplies and personnel are such as to require assistance from state or federal agencies.

Why Plans Fail

Although many jurisdictions have adopted formal disaster plans, local governments often fail to improve their plans, even after a major disaster has occurred. Local governments also fail to adequately and effectively plan for disaster response. These failures can often be attributed to lack of relevant experience with disaster response, failure to study lessons learned, failure to commit to carrying out a disaster planning program, or performing the wrong kind of planning.

Many communities lack the experience to deal with disaster response because their public officials are not involved in enough disasters to gain personal experience. There is often the misconception that responding to a disaster is the same as responding to any other emergency, only on a grander scale.

Another impediment to planning is that although an individual may have been involved in disaster response, it is difficult to view the disaster from the perspective of other organizations. Postdisaster critiques often turn out to be justifications of actions taken, rather than impartial, objective assessments of problems and mistakes.

Disaster planning must have the support of the entire community if it is to be successful. Lack of public awareness can often undercut the community's efforts to plan. Key officials often neglect to read emergency plans. Even after a plan is written, it is often not properly exercised, often resulting in failure of the plan during a true disaster.

Planning As a Blueprint

A disaster plan should serve as a community's blueprint for initiating, managing, and performing operations that will most likely extend beyond the scope of functions carried out in normal day-to-day operations. The disaster plan should serve to coordinate the activities, logistics, and resources involved in disaster response. Plans typically seek to establish the various sectors that need to be implemented in the event of a disaster.

A comprehensive emergency management strategy includes four phases that work in a pattern. The four phases are mitigation, preparedness, response, and recovery. The phases are not linear in nature, but are more cyclical, as illustrated in Figure 1.1.

The goal of prevention is to avert accidents and emergencies. This is not always possible, however, thus creating the need for preparedess, the component of planning in which steps are taken to ensure that all individuals and entities to be involved in a disaster response ready themselves to perform during an emergency.

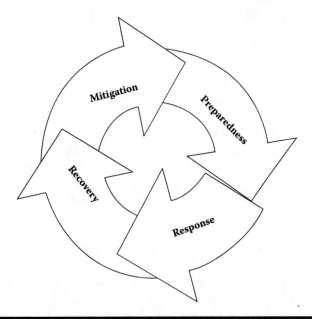

Figure 1.1   The four phases of comprehensive emergency management.

The key to preparedness is to ensure that an adequate level of resources is available to save lives and minimize property damage.

The next stage in disaster planning is response, which is the initiation of activities to save lives and protect property. The final step is recovery. It is in the recovery stage that residents and government try to resume business as usual.

Again, these phases are more cyclic in nature than they are linear as the process is always ongoing and continuous.

The Disaster Cycle and Planning

Without planning, emergency operations can suffer from a multitude of problems, which could lead to serious consequences and even death. Therefore, it is imperative to have a system that enables all participants involved in an emergency response to work together. An integrated emergency management system does just that.

An integrated and comprehensive emergency management system is a conceptual framework that increases emergency management capability by using a structured approach to planning and response. To have this increased capability, it is important to establish networking, coordination, interoperability, partnerships, and creative thinking about resource shortfalls during the planning process before an emergency occurs.

An integrated and comprehensive emergency management system should:

- Address all hazards that threaten a community
- Be useful in all phases of emergency management
- Seek resources from any and all sources that are appropriate
- Knit together all partnerships and participants to achieve a mutual goal

The goals of emergency management are to save lives, prevent injuries, and protect property and the environment. Participants in emergency management should include local, state, and federal government; private sector entities (e.g., nonprofit organizations, businesses, and industry); and private citizens.

The role of local government in emergency management is not just limited to response. Local government must ensure the safety of its citizens and acquire knowledge of the threats to the community and the resources required to meet those threats. Some of the measures that local government should take to prepare for an emergency include:

- Ensuring that the emergency operations plan (EOP) is developed, trained, exercised, and maintained
- Developing mutual aid agreements, memoranda of understanding (MOUs), and stand-by contracts for critical emergency resources
- Communicating with the public about potential hazards and how to prepare for them

State government has legal authorities for emergency response and recovery, and serves as the point of contact between local and federal governments. In addition, state governments may require that local jurisdictions develop and submit emergency plans for state review and incorporation into the state plan. State governments also have the authority to activate National Guard resources in response to emergencies and develop emergency mutual aid compacts (EMACs) with other states.

To assist state and local governments in recovering from emergencies, the federal government has legal authorities, fiscal resources, research capabilities, technical information and services, and specialized personnel.

Although government agencies are responsible for protecting the public, they need the help of private citizens in all facets of emergency management to be successful. Nonprofit organizations, such as the American Red Cross, offer critical resources in emergencies.

Private citizens can assist by reducing hazards in and around their communities, preparing disaster supply kits, and monitoring emergency communications carefully. Citizens can also volunteer with established organizations and pursue training in emergency preparedness and response.

Resources for an integrated emergency management system include both equipment resources and personnel resources. Personnel resources include:

- Elected and appointed officials
- Emergency program managers
- Emergency operations staff
- Police and fire departments
- Other local service providers, such as public works, transportation, and public health
- Voluntary organizations (e.g., American Red Cross, Salvation Army)

An integrated and comprehensive emergency management system brings all of these personnel resources together through:

Planning—Involving all of these key players in the planning process ensures that all roles and responsibilities are clearly defined;

Direction—All parties must have direction clearly defined in the EOP to reduce freelancing and ensure that all response activities are handled according to established policies and procedures;

Coordination—Working together during the planning process develops relationships and promotes teamwork during a response;

Clearly defined roles and responsibilities—Developing and agreeing to roles and responsibilities during the planning process reduces redundancy and ensures the most efficient use of resources during an emergency.

Further, an integrated and comprehensive emergency management program:

- Examines potential emergencies and disasters based on the risks posed by likely hazards
- Develops and implements programs aimed at reducing the impact of these events on the community
- Prepares for hazards that cannot be eliminated
- Prescribes actions required to deal with the consequences of actual events

As discussed previously, comprehensive emergency management activities are divided into four phases:

Mitigation—Taking sustained actions to reduce or eliminate long-term risk to people and property from hazards;

Preparedness—Building the emergency management function to respond effectively to, and recover from, any hazard;

Response—Conducting emergency operations to save lives and property by taking actions to reduce hazards; evacuating potential victims; providing food, water, and medical care to those in need; and completing emergency repairs to restore critical public services;

Recovery—Rebuilding communities so that individuals, businesses, and governments can function on their own, return to normal life, and protect against future hazards.

Following an emergency, lessons learned are used to mitigate, prepare, and respond better. As the plan is revised based on lessons learned, the cycle repeats. Each of these phases is described next.

Mitigation

As the costs of disasters continue to rise, it is necessary to take sustained action to reduce or eliminate the long-term risk to people and property from hazards and their effects. These sustained actions are known as mitigation.

Mitigation should be a continued activity that is integrated with each of the other phases of emergency management to employ a long-range, community-based approach to disasters. The goals of mitigation activities are to protect people and structures, and reduce the cost of response and recovery.

Mitigation is accomplished in conjunction with a hazard analysis, which helps to identify what events can occur in and around the community; the likelihood that an event will occur; and the consequences of the event in terms of casualties, destruction, disruption to critical services, and costs of recovery. To be successful, mitigation measures must be developed into an overall mitigation strategy that considers ways to reduce hazard losses together with the overall risk from specific hazards and other community goals.

Because it is not possible to mitigate completely against every hazard, preparedness measures can help to reduce the impact of those hazards by taking certain actions before an emergency occurs.

Preparedness

Preparedness involves all of the players in the integrated emergency management system and includes activities, such as:

■ Developing, training, and exercising the EOP
■ Recruiting, assigning, and training staff who can assist in key areas of response operations
■ Identifying resources and supplies that might be required in an emergency
■ Designating facilities for emergency use

Response

Response begins when an emergency is imminent or immediately after an event occurs. Response encompasses all activities taken to save lives and reduce damage from the event and includes providing emergency assistance to victims, completing emergency repairs critical to infrastructure, and ensuring the continuity of critical services.

One of the first response activities should be to conduct a situation assessment. To fulfill this task, responders must conduct an immediate rapid assessment of the situation. Rapid assessment includes all immediate response activities that are directly related to determining initial lifesaving and life-sustaining needs and identifying imminent hazards. Coordinated and timely assessments enable local government to prioritize response activities, allocate scarce resources, and request additional assistance from mutual aid partners.

Recovery

The final phase in the emergency management cycle is recovery, the goal of which is to return the community's systems and activities to normal. Some recovery operations may be concurrent with response efforts.

Recovery from a disaster is unique to each community and disaster. Short-term recovery is an extension of the response phase, in which basic functions and services are restored. After short-term recovery, the community must rebuild. Considerations for long-term recovery include:

■ Applying for federal assistance
■ Keeping the public informed on the rebuilding process
■ Taking mitigation measures to ensure against future disaster damage
■ Collecting and distributing donations

- Building partnerships with business and industry for needed resources
- Taking care of environmental concerns
- Meeting the needs of victims
- Taking public health measures to protect against diseases and contamination
- Rebuilding bridges, roads, and other elements of the community's infrastructure

Recovery also involves taking steps necessary to reopen damaged businesses, reemploy workers, and other measures required to return the community to its pre-emergency status.

A Brief History of Emergency Management

Throughout its history, the United States has faced many disasters—from natural disasters to hazardous materials releases to terrorist attacks. Although all disasters are local, the federal government will provide support—personnel, equipment, resources, and funding—when local capabilities and capacities are exceeded.

Before the 1970s, various federal agencies and programs provided disaster relief services. At one point, more than one hundred federal agencies handled disaster and emergencies. However, until four devastating hurricanes struck in the 1970s, the need for improved disaster coordination at the federal level did not happen. Following are the four hurricanes:

- Hurricane Agnes (1972)—This hurricane caused significant East Coast flooding and $2 billion in damages.
- Hurricane Eloise (1975)—This hurricane caused $200 million in damages and 76 fatalities.
- Hurricanes David and Frederick (1979)—These hurricanes were among the deadliest ever seen in the Caribbean, with Frederick causing $2.2 billion in damages.

These disasters and the need for disaster preparedness nationwide pushed the Carter administration to establish the Federal Emergency Management Agency (FEMA) to coordinate all disaster relief efforts at the federal level.

Authorities and Directives

After creating FEMA, the federal government continued to refine disaster response by creating several acts and directives. Some, such as the Robert T. Stafford Act, were created as general, all-hazards guidelines, whereas others, such as the Post-Katrina Emergency Management Reform Act, were born from a specific event.

Robert T. Stafford Act

To bring a more orderly and systemic means of federal natural disaster assistance for state and local governments in carrying out their responsibilities to aid citizens, the Robert T. Stafford Disaster Relief and Emergency Assistance Act (Public Law 93-288) was passed in 1988. This act describes the programs and processes by which the federal government provides coordination and support to state and local governments, tribal nations, eligible nongovernment organizations (NGOs), and individuals affected by a declared major disaster or emergency. The Stafford Act covers all hazards, including natural disasters and terrorist events.

The Stafford Act was amended several times, most recently with the Post-Katrina Emergency Management Reform Act (PKEMRA) in 2006 (Public Law 109-295) to address the issues and challenges that arose during the response to Hurricane Katrina (see "Post-Katrina Emergency Management Reform Act" section for more information on PKEMRA).

Homeland Security Act of 2002

Title I of the Homeland Security Act of 2002 established the Department of Homeland Security (DHS), with the mission of:

- Preventing terrorist attacks within the United States
- Reducing the vulnerability of the United States to terrorism
- Minimizing the damage, and assisting in the recovery, from terrorist attacks that do occur within the United States
- Carrying out all functions of entities transferred to the department, including by acting as a focal point regarding natural and manmade crises and emergency planning

The act organized FEMA into DHS, with a direct line of report between the administrator and the secretary, but it also kept it separate from the new Preparedness Directorate. The Preparedness Directorate consolidated preparedness assets from across the department. It facilitated grants and oversaw nationwide preparedness efforts supporting first-responder training, citizen awareness, public health, infrastructure, and cyber security.

Post-Katrina Emergency Management Reform Act

The Post-Katrina Emergency Management Reform Act of 2006 clarified and modified the Homeland Security Act of 2002 with respect to the organizational structure, authorities, and responsibilities of FEMA and the FEMA administrator. Among other changes, PKEMRA transferred a significant portion of the DHS Preparedness Directorate into FEMA, notably:

- United States Fire Administration
- Office of Grants and Training
- Chemical Stockpile Emergency Preparedness Division
- Radiological Emergency Preparedness Program
- Office of National Capital Region Coordination

Presidential Decision Directives

In 1995, the Clinton administration issued Presidential Decision Directive 39 (PDD-39), "U.S. Policy on Counterterrorism" in response to the worst terrorist act on U.S. soil—the bombing of the Alfred P. Murrah Federal Building in Oklahoma City. PDD-39, built on prior directives, outlined three key elements of a national counterterrorism strategy:

1. Reduce vulnerabilities to terrorist attacks and prevent and deter terrorist acts before they occur (threat/vulnerability management)
2. Respond to terrorist acts that occur, end the crisis or deny terrorists their objectives, and apprehend and punish terrorists (crisis management)
3. Manage the consequences of terrorist acts, including restoring essential government services and providing emergency relief, to protect public health and safety (consequence management)

The directive elaborated on specific roles and responsibilities for several federal agencies with respect to each element of the strategy. For example, PDD-39 gave the Federal Bureau of Investigation (FBI) lead agency responsibility for crisis management, and FEMA similar responsibility for consequence management. Reflecting the need for greater interagency coordination, PDD-39 also directed the National Security Council to coordinate interagency terrorism policy issues and to ensure implementation of federal counterterrorism policy and strategy.

Homeland Security Presidential Directives

Following the terrorist attacks of September 11, 2001, the Bush administration issued several Homeland Security Presidential Directives (HSPDs), either augmenting or replacing the PDDs established by the Clinton administration. Among several, the two most notable HSPDs concerning preparedness and response are HSPD-5 and HSPD-8.

HSPD-5, Management of Domestic Incidents

Issued by the White House on February 28, 2003, HSPD-5 established a single, comprehensive national incident management system. It also designated the secretary of Homeland Security as the principal federal official for domestic incident management and recognizes the statutory authorities of the attorney general, secretary of defense, and secretary of state. In addition, HSPD-5 directed the heads of all federal departments and agencies to provide their full and prompt cooperation, resources, and support, as appropriate and consistent with their own responsibilities for protecting national security, to the secretary of Homeland Security, attorney general, secretary of defense, and secretary of state in the exercise of leadership responsibilities and missions assigned.

HSPD-8, National Preparedness

Issued by the White House on December 17, 2003, HSPD-8 established policies to strengthen the preparedness of the United States to prevent and respond to threatened or actual domestic terrorist attacks, major disasters, and other emergencies by requiring a national domestic all-hazards preparedness goal, establishing mechanisms for improved delivery of federal preparedness assistance to state, local, and tribal governments, and outlining actions to strengthen preparedness capabilities of federal, state, local, and tribal entities. Annex 1, Integrated Planning System, published in January 2009, established a standard and comprehensive approach to national planning.

Other Policy References

HSPD-5 and HSPD-8 also helped establish several important policy references around homeland security and emergency management, including the following:

- National Incident Management System (NIMS), December 2008, provides a systematic, proactive approach to guide departments and agencies at all levels of government, NGOs, and the private sector to work seamlessly to prevent, protect against, respond to, recover from, and mitigate the effects of incidents, regardless of cause, size, location, or complexity, in order to reduce the loss of life and property and harm to the environment.
- National Infrastructure Protection Plan (NIPP), January 2009, establishes a risk management framework for the nation's unified national approach to critical infrastructure and key resource protection.
- National Preparedness Guidelines, September 2007, finalizes development of the National Preparedness Goal and its related preparedness tools as mandated in HSPD-8. The guidelines consist of four elements: the National Preparedness Vision, the National Planning Scenarios, the Target Capabilities List, and the Universal Task List.

Response Plans

For nearly a decade, a progression of response plans—Federal Response Plan, National Response Plan, and National Response Framework—were written to address lessons learned from actual and potential disasters, as well as changes in statutory and policy directives, with the intended aim of improving the nation's preparedness and response coordination.

The creation of the Federal Response Plan (FRP) was driven by the PDD-39, which itself was developed following the Oklahoma City bombing of the Murrah Federal Building. The National Response Plan (NRP) superseded the FRP, following the terrorist attacks of September 11, 2001. Finally, the National Response Framework (NRF) superseded the NRP after the devastation of Hurricane Katrina. Creation of both the NRP and the NRF was driven by HSPD-5.

These federal plans are supported by emergency support functions (ESFs) annexes that describe the missions, policies, structures, and responsibilities of federal agencies for coordinating resource and programmatic support. The ESFs provide the structure for coordinating federal interagency support for a federal response to an incident. They are mechanisms for grouping functions most frequently used to provide federal support.

The plans are guidelines, not requirements for the states in the beginning, but states are supposed to follow federal plans to get funding. Some states follow the plan, others do not. The development of these plans was based on several premises:

A basic premise of all the plans is that the state is FEMA's primary client. The response doctrine is rooted in America's federal system and the Constitution's division of responsibilities between federal and state governments.

Another premise is that incidents are handled at the lowest jurisdictional level possible. In the vast majority of incidents, state and local resources and interstate mutual aid will provide the first line of emergency response and incident management support. When state resources and capabilities are overwhelmed, governors may request federal assistance. That strategy provides the framework for federal interaction with local, tribal, state, territory, commonwealth, and private-sector and nongovernmental entities in the context of domestic incident management.

A third premise is that mass care is traditionally a community response. NGOs, such as the American Red Cross and the Salvation Army, and other faith-based and community-based organizations have traditionally provided mass care services to communities during disasters.

All of these plans focus upon all-hazard emergencies: natural disasters, technological emergencies (such as hazardous material releases), and acts of terrorism.

Federal Response Plan

The FRP established a new process and structure for the systematic, coordinated, and effective delivery of federal assistance to address the consequences of any major disaster or emergency declared under the Stafford Act. The plan organized the types of federal response assistance that a state is most likely to need under twelve ESFs. It also described the process and methodology for implementing and managing federal recovery and mitigation programs and support/technical services.

The FRP provided a focus for interagency and intergovernmental emergency preparedness, planning, training, exercising, coordination, and information exchange, serving as the foundation for the development of detailed supplemental plans and procedures to implement federal response and recovery activities rapidly and efficiently.

The FRP applied to a major disaster or emergency as defined under the Stafford Act for which the president determines that federal assistance is needed to supplement state and local efforts and capabilities. The FRP covered the full range of complex and constantly changing requirements following a disaster: saving lives, protecting property, and meeting basic human needs (response); restoring the disaster-affected area (recovery); and reducing vulnerability to future disasters (mitigation). The FRP did not specifically address long-term reconstruction and redevelopment.

The FRP engaged twenty-two federal agencies, plus the American Red Cross, to assist states with disaster preparedness and response. The FRP augmented other response such as the National Contingency Plan (NCP) for oil and hazardous materials spills and the Federal Radiological Emergency Response Plan (FRERP).

Under the FRP, the federal government and the American Red Cross shared responsibility for sheltering victims, organizing feeding operations, providing emergency first aid at designated sites, collecting and providing information on victims to family members, and coordinating bulk distribution of emergency relief items. As the primary agency for mass care under ESF 6, the American Red Cross coordinated federal mass care assistance in support of state and local mass care efforts. The American Red Cross was the only NGO signatory to the FRP. Other NGOs became formally involved in later plans.

While the FRP was revised in 2003 in response to the terrorist attacks of September 11, 2001, it was acknowledged that further revision was needed to include long-term recovery activities (e.g., housing) as well as response services.

National Response Plan

In 2003, President George W. Bush directed DHS, through HSPD-5, to develop a new national response plan to align federal coordination structures, capabilities, and resources to form a unified, all-discipline, and all-hazards approach to domestic incident management. This approach eliminated critical seams and tied together a complete spectrum of incident management activities to include the prevention of, preparedness for, response to, and recovery from terrorism, major natural disasters, and other major emergencies.

In December 2004, Secretary Tom Ridge issued the NRP, which described how to improve coordination among federal, state, local, and tribal organizations to help save lives and protect America's communities by increasing the speed, effectiveness, and efficiency of incident management.

As noted by the preface, Secretary Ridge acknowledged:

> Implementation of the plan and its supporting protocols requires extensive cooperation, collaboration, and information-sharing across jurisdictions, as well as between the government and the private sector at all levels.

The NRP was built on the template of NIMS, which provides a consistent doctrinal framework for incident management at all jurisdictional levels, regardless of the cause, size, or complexity of the incident. It superseded other response plans, namely,

- Federal Response Plan
- United States Government Interagency Domestic Terrorism Concept of Operations Plan
- Federal Radiological Emergency Response Plan

The NRP and its coordinating structures and protocols provided mechanisms for coordination and implementation of a wide variety of incident management and emergency response activities (see Figure 1.2). The NRP was an integration of the state, local, and federal assets. Included in these coordinating activities were federal support to local, tribal, and state authorities; interaction with nongovernmental, private-donor, and private-sector organizations; and the coordinated, direct exercise of federal authorities, when appropriate.

Whereas the FRP addressed response activities only, the NRP included response and recovery, as well as a need for long-term recovery activities that was recognized following Hurricane Katrina. The NRP was considered by the emergency management community not to be a plan but rather to set boundaries for hierarchical framework for planning.

The NRP also established a new term of reference for disasters—incidents of national significance—to address potential acts of terrorsim. Under the authority of the secretary of Homeland Security, federal response to an incident of national significance could include:

- A federal department or agency, responding under its own authorities, requests DHS assistance

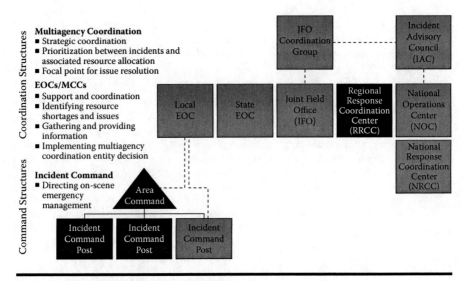

Figure 1.2   Coordination and command structures in the NRP.

■ Resources of state and local authorities are overwhelmed
   – Stafford Act for major disasters or emergencies
   – Other catastrophic incidents
■ More than one federal department or agency is involved
   – Credible threats or indications of imminent terrorist attack
   – Threats/incidents related to high-profile, large-scale events

Given its sweeping changes, the NRP included its own implementation schedule (see Figure 1.3).

The plan addressed the full spectrum of activities related to domestic incident management, including prevention, preparedness, response, and recovery actions. The NRP focused on those activities that are directly related to an evolving incident or potential incident rather than steady-state preparedness or readiness activities conducted in the absence of a specific threat or hazard.

Additionally, since incidents of national significance typically resulted in impacts far beyond the immediate or initial incident area, the NRP provided a framework to enable the management of cascading impacts and multiple incidents as well as the prevention of and preparation for subsequent events. Examples of incident management actions from a national perspective include:

■ Increasing nationwide public awareness
■ Assessing trends that point to potential terrorist activity
■ Elevating the national Homeland Security Advisory System alert condition and coordinating protective measures across jurisdictions

- **Phase I—Transitional Period (0 to 60 days):** This 60-day timeframe is intended to provide a transitional period for departments and agencies and other organizations to modify training, designate staffing of NRP organizational elements, and become familiar with NRP structures, processes, and protocols.

- **Phase II—Plan Modification (60 to 120 days):** This second 60-day timeframe is intended to provide departments and agencies the opportunity to modify existing federal interagency plans to align with the NRP and conduct necessary training.

- **Phase III—Initial Implementation and Testing (120 days to 1 year):** Four months after its issuance, the NRP is to be fully implemented, and the INRP, FRP, CONPLAN, and the FRERP are superseded. Other existing plans remain in effect, modified to align with the NRP. During this timeframe, the Department of Homeland Security (DHS) will conduct systematic assessments of NRP coordinating structures, processes, and protocols implemented for actual incidents of national significance (defined on page 4 of the NRP), national-level homeland security exercises, and national special security events (NSSEs). These assessments gauge the plan's effectiveness in meeting specific objectives outlined in Homeland Security Presidential Directive-5 (HSPD-5). At the end of this period, DHS will conduct a one-year review to assess the implementation on process and make recommendations to the secretary on necessary NRP revisions. Following this initial review, the NRP will begin a deliberate four-year review and reissuance cycle.

Figure 1.3   The three phases of the National Response Plan.

- Increasing countermeasures such as inspections, surveillance, security, counterintelligence, and infrastructure protection
- Conducting public health surveillance and assessment processes and, where appropriate, conducting a wide range of prevention measures to include, but not be limited to, immunizations
- Providing immediate and long-term public health and medical response assets
- Coordinating federal support to state, local, and tribal authorities in the aftermath of an incident
- Providing strategies for coordination of federal resources required to handle subsequent events
- Restoring public confidence after a terrorist attack
- Enabling immediate recovery activities, as well as addressing long-term consequences in the impacted area

On August 30, 2005, Secretary Michael Chertoff invoked the NRP the day after Hurricane Katrina hit the Gulf Coast. By so doing, Secretary Chertoff assumed the leadership role triggered by the law to bear primary responsibility to manage said crisis. Almost a month later, in advance of the landfall of Hurricane Rita, Secretary Chertoff declared the storm an incident of national significance and put preparations in place in the gulf region of Texas.

Because of the lengthy implementation schedule, the increased level of coordination did not sufficiently materialize. This situation became severely problematic when Hurricane Katrina roared into the Gulf of Mexico, then made landfall in Louisiana. Hurricane Katrina caused severe destruction along the Gulf coast. The most severe loss of life and property damage occurred in New Orleans, Louisiana, which flooded as the levee system catastrophically failed, in many cases hours after the storm had moved inland.

Following Hurricane Katrina, the plan was updated on May 25, 2006. The notice of change stated the update "emerged from organizational changes within DHS, as well as the experience of responding to Hurricanes Katrina, Wilma, and Rita in 2005."

National Response Framework

Published in January 2008, the NRF was developed to address the requirements of PKEMRA. It is a framework that guides local, state, and federal entities enabling all response partners to prepare for and provide a unified national response to disasters and emergencies. This framework establishes a comprehensive, national, all-hazards approach to domestic incident approach.

As identified by DHS, the NRF

> presents the guiding principles that enable all response partners to pre-
> pare for and provide a unified national response to disasters and emer-
> gencies—from the smallest incident to the largest catastrophe. This
> important document establishes a comprehensive, national, all-hazards
> approach to domestic incident response. The Framework defines the
> key principles, roles, and structures that organize the way we respond
> as a Nation. It describes how communities, tribes, States, the Federal
> Government, and private-sector and nongovernmental partners apply
> these principles for a coordinated, effective national response. It also
> identifies special circumstances where the Federal Government exercises
> a larger role, including incidents where Federal interests are involved
> and catastrophic incidents where a State would require significant sup-
> port. The Framework enables first responders, decision-makers, and
> supporting entities to provide a unified national response.

An underlying basis of the NRF is a set of key principles:

Engaged partnership—Leaders at all levels must communicate and actively support engaged partnerships by developing shared goals and aligning capabilities so that no one is overwhelmed in times of crisis.

Tiered response—Incidents must be managed at the lowest possible jurisdictional level and supported by additional capabilities when needed.

Scalable, flexible, and adaptable operational capabilities—As incidents change in
size, scope, and complexity, the response must adapt to meet requirements.

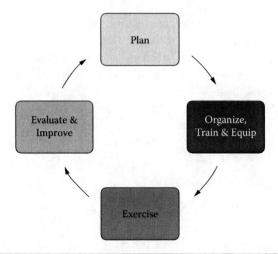

Figure 1.4   Preparedness life cycle of the NRF.

Unity of effort through unified command—Effective unified command is indispensable to response activities and requires a clear understanding of the roles and responsibilities of each participating organization.

Readiness to act—Effective response requires readiness to act balanced with an understanding of risk. From individuals, households, and communities to local, tribal, state, and federal governments, national response depends on the instinct and ability to act.

Because of the confusion brought about by the term of reference—incident of national significance—in the NFR, the NRP eliminated this term.

An important concept presented in the NRF included the preparedness life cycle, which represents a systemic approach to build the right capabilities for the nation in response to all hazards. The preparedness life cycle (see Figure 1.4):

- Introduces the National Planning System
- Defines response organization
- Requires training
- Advocates interoperability and typing of equipment;
- Emphasizes exercising with broad-based participation
- Describes the process for continuous evaluation and improvement

The NRF establishes fifteen emergency support functions:

- ESF #1—Transportation
- ESF #2—Communications

- ESF #3—Public Works and Engineering
- ESF #4—Firefighting
- ESF #5—Emergency Management
- ESF #6—Mass Care, Emergency Assistance, Housing and Human Services
- ESF #7—Logistics Management and Resource Support
- ESF #8—Public Health and Medical Services
- ESF #9—Search and Rescue
- ESF #10—Oil and Hazardous Materials Response
- ESF #11—Agriculture and Natural Resources
- ESF #12—Energy
- ESF #13—Public Safety and Security
- ESF #14—Long-Term Community Recovery
- ESF #15—External Affairs

It also includes several support annexes and incident annexes to help improve coordination:

- Support Annexes
  - Critical Infrastructure and Key Resources
  - Financial Management
  - International Coordination
  - Private Sector Coordination
  - Public Affairs
  - Tribal Relations
  - Volunteer and Donations Management
  - Worker Safety and Health
- Incident Annexes
  - Biological Incident
  - Catastrophic Incident
  - Cyber Incident
  - Food and Agriculture Incident
  - Mass Evacuation Incident
  - Nuclear/Radiological Incident
  - Terrorism Incident Law Enforcement and Investigation

To promote awareness and education of the NRF, FEMA developed an independent study training course, IS-800, An Introduction to the NRF, which is available free of charge. FEMA continues to develop other general orientation courses for ESFs and the Support and Incident Annexes though its on-line study program at the Emergency Management Institute.

Emergency Support Functions

FRP, NRP, and NRF use emergency support functions (ESFs) as a means to provide the interagency staff to support federal response operations of the National Response Coordination Center (NRCC), the Regional Response Coordination Center (RRCC), and the Joint Field Office (JFO). Depending on the incident, deployed assets of the ESFs may also participate in the staffing of the Incident Command Post. Under the NRP, each ESF is structured to provide optimal support for evolving incident management requirements.

ESFs may be activated for Stafford Act and non-Stafford Act implementation of the NRP (although some incidents of national significance may not require ESF activations). ESF funding for non-Stafford Act situations will be accomplished using NRP federal-to-federal support mechanisms and will vary based on the incident.

Within the NRP, each ESF annex identifies the ESF coordinator and the primary and support agencies pertinent to the ESF. Several ESFs incorporate multiple components, with primary agencies designated for each component to ensure seamless integration of and transition between preparedness, prevention, response, recovery, and mitigation activities. ESFs with multiple primary agencies designate an ESF coordinator for the purposes of preincident planning and coordination.

A federal agency designated as an ESF primary agency serves as a federal executive agent under the federal coordinating officer (FCO; or federal resource coordinator for non-Stafford Act incidents) to accomplish the ESF mission. When an ESF is activated in response to an incident of national significance, the primary agency is responsible for:

- Orchestrating federal support within their functional area for an affected state
- Providing staff for the operations functions at fixed and field facilities
- Notifying and requesting assistance from support agencies
- Managing mission assignments and coordinating with support agencies, as well as appropriate state agencies
- Working with appropriate private-sector organizations to maximize use of all available resources
- Supporting and keeping other ESFs and organizational elements informed of ESF operational priorities and activities
- Planning for short-term and long-term incident management and recovery operations

ESF Support Agencies

When an ESF is activated in response to an incident of national significance, support agencies are responsible for:

- Conducting operations, when requested by DHS or the designated ESF primary agency, using their own authorities, subject-matter experts, capabilities, or resources
- Participating in planning for short-term and long-term incident management and recovery operations and the development of supporting operational plans, procedures, checklists, or other job aids, in concert with existing first-responder standards
- Assisting in the conduct of situational assessments
- Furnishing available personnel, equipment, or other resource support as requested by DHS or the ESF primary agency
- Providing input to periodic readiness assessments
- Identifying new equipment or capabilities required to prevent or respond to new or emerging threats and hazards, or to improve the ability to address existing threats
- Nominating new technologies to DHS for review and evaluation that have the potential to improve performance within or across functional areas
- Providing information or intelligence regarding their agency's area of expertise

Conclusion

Government at all levels has the basic responsibility for protecting its citizens. In the case of natural, technological, or manmade hazards, the best way to protect the public is by implementing an integrated emergency management system that incorporates all potential players in a response through all phases of emergency management.

As technology continues to advance, manmade disasters will likely reach a magnitude never thought possible. The basic process of disaster planning is only a small part of the process. Equally important is identifying who is responsible for developing or revamping a community's disaster plan. These are the topics to be covered in this book.

Chapter 2

Assessing Vulnerabilities

James Peerenboom and Ron Fisher

Contents

Introduction

The Homeland Security Act of 2002 provides the primary authority for the overall homeland security mission. This act charged the Department of Homeland Security (DHS) with primary responsibility for developing a comprehensive national plan to secure critical infrastructure and key resources (CIKR) and recommend "the measures necessary to protect the key resources and critical infrastructure of the United States." This comprehensive plan is the National Infrastructure Protection Plan (NIPP), first published by the DHS in June 2006. As defined in the 2009 NIPP, critical infrastructure are the systems and assets, whether physical or virtual, so vital that the incapacity or destruction of such may have a debilitating impact on the security, economy, public health or safety, environment, or any combination of these matters, across any federal, state, regional, territorial, or local jurisdiction. Key resources are publicly or privately controlled resources essential to the minimal operations of the economy and government. The NIPP provides the unifying structure for integrating a wide range of efforts for the protection of CIKR into a single national program.

Homeland Security Presidential Directive 7 (HSPD-7), Critical Infrastructure Identification, Prioritization, and Protection, was established as a national policy for federal departments and agencies to identify and prioritize United States CIKR and to protect them from terrorist attacks. The NIPP provided the follow-up plan to implement HSPD-7. The NIPP called out the need to conduct risk assessments to deter threats, mitigate vulnerabilities, and minimize consequences.

Vulnerability Assessment

Vulnerability assessment methodologies are generally intended to identify any weakness that can be exploited by an adversary to gain unauthorized access to or to disrupt an asset, facility, or system. Terrorism is often the primary focus; however, vulnerabilities can take an all-hazards approach. Vulnerabilities can result from, but are not limited to, the following:

- Asset, building, site, or system characteristics
- Equipment properties
- Personal behavior
- Operational and personnel practices
- Security weaknesses (physical and cyber)

Vulnerability assessment methodologies can be characterized in terms of four assessment elements—physical, cyber, operations security, and interdependencies. Each is briefly described next.

A *physical security assessment* typically evaluates the physical security systems in place or planned at a site, including access controls, barriers, locks and keys, badges and passes, intrusion detection devices and associated alarm reporting and display, closed-circuit television (CCTV) (assessment and surveillance), communications equipment (telephone, two-way radio, intercom, cellular), lighting (interior and exterior), power sources (line, battery, generator), inventory control, postings (signs), security system wiring, and protective force. These systems are generally reviewed for design, installation, operation, maintenance, and testing. It may also include an evaluation of sites housing critical equipment or information assets or networks dedicated to the operation of the physical systems.

A *cyber security assessment* evaluates the security features of the information network(s) associated with an organization's critical information systems. This could include an examination of network topology and connectivity, principal information assets, interface and communications protocols, function and linkage of major software and hardware components (especially those associated with information security such as intrusion detectors), and policies and procedures that govern security features of the network. It may also include internal and external scanning for vulnerabilities (penetration testing).

Operations security (OPSEC) is the systematic process of denying potential adversaries information about capabilities and intentions of the host organization. This is accomplished by identifying, controlling, and protecting generally nonsensitive activities concerning planning and execution of sensitive activities. An OPSEC assessment typically reviews the processes and practices employed for denying adversary access to sensitive and nonsensitive information that might inappropriately aid or abet any individual's or organization's disproportionate influence over system operation. This should include a review of security training and awareness programs, a review of personnel policies and procedures, discussions with key staff, and tours of appropriate principal facilities. It should also include a review of information that may be available through public access (e.g., the Internet).

Infrastructure interdependencies refers to the physical and electronic (cyber) linkages within and among our nation's critical infrastructures (i.e., within and among the thirteen critical infrastructure sectors and five key asset categories). An interdependencies assessment typically identifies the direct infrastructure linkages between and among both the internal infrastructures at a site as well as the linkages to external infrastructures outside the site. The process of identifying and analyzing these linkages requires a detailed understanding of how the components of each infrastructure and their associated functions or activities depend on, or are supported by, each of the other infrastructures. For example, a supervisory control and data acquisition (SCADA) system that operates a natural gas pipeline depends on the local electric power and telecommunications infrastructures to function. The failure of a separate, external infrastructure could prevent the SCADA system from operating, thus impacting natural gas deliveries to or within a system. Interdependencies can create subtle interactions and feedback mechanisms that often lead to unintended behaviors and consequences, including the potential disruption of critical infrastructures.

Critical Infrastructure and Key Resources (CIKR)

Attacks on CIKR could significantly disrupt the functioning of government and business alike and produce cascading effects far beyond the targeted sector and physical location of the incident (see Figure 2.1). Direct terrorist attacks and natural, manmade, or technological hazards could produce catastrophic losses in terms of human casualties, property destruction, and economic effects, as well as profound damage to public morale and confidence. Finally, attacks using components of the nation's CIKR as weapons of mass destruction could have even more devastating physical and psychological consequences. CIKR sectors are described next.

Agriculture and Food

The Agriculture and Food Sector has the capacity to feed and clothe people well beyond the boundaries of the nation. The sector is almost entirely under private ownership and is composed of an estimated 2.1 million farms, approximately 880,500 firms, and over 1 million facilities. This sector accounts for roughly one-fifth of the nation's economic activity and is overseen at the federal level by the U.S. Department of Agriculture (USDA) and the Department of Health and Human Services (HHS), and Food and Drug Administration (FDA).

Banking and Finance

The Banking and Finance Sector, the backbone of the world economy, is a large and diverse sector primarily owned and operated by private entities. In 2007, the sector accounted for more than 8 percent of the U.S. gross domestic product.

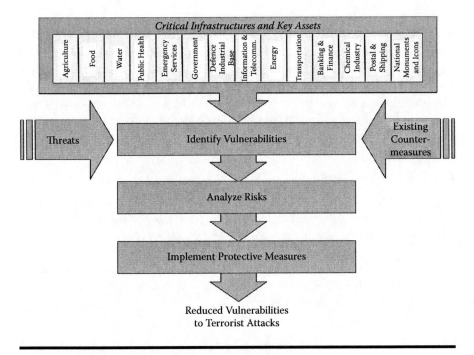

Figure 2.1   Protective security process for CIKR.

Chemical

The Chemical Sector is an integral component of the U.S. economy, employing nearly 1 million people, and earning revenues of more than $637 billion per year. This sector can be divided into five main segments, based on the end product produced: basic chemicals, specialty chemicals, agricultural chemicals, pharmaceuticals, and consumer products.

Commercial

Facilities associated with the Commercial Facilities Sector operate on the principle of open public access, meaning that the general public can move freely throughout these facilities without the deterrent of highly visible security barriers. The majority of the facilities in this sector are privately owned and operated, with minimal interaction with the federal government and other regulatory entities.

Communications

The Communications Sector is an integral component of the U.S. economy as it underlies the operations of all businesses, public safety organizations, and government. Over twenty-five years, the sector has evolved from predominantly a provider of voice services into a diverse, competitive, and interconnected industry using terrestrial, satellite, and wireless transmission systems. The transmission of these services has become interconnected; satellite, wireless, and wireline providers depend on one another to carry and terminate their traffic, and companies routinely share facilities and technology to ensure interoperability. A majority of the Communications Sector is privately owned, requiring the DHS to work closely with the private sector and its industry associations to identify infrastructure, assess and prioritize risks, develop protective programs, and measure program effectiveness.

Critical Manufacturing

The Critical Manufacturing Sector is crucial to the economic prosperity and continuity of the United States. U.S. manufacturers design, produce, and distribute products that provide more than one of every eight dollars of the U.S. gross domestic product and employ more than 10 percent of the nation's workforce. A direct attack on or disruption of certain elements of the manufacturing industry could disrupt essential functions at the national level and across multiple other critical infrastructure and key resources sectors.

Dams

The Dams Sector comprises the assets, systems, networks, and functions related to dam projects, navigation locks, levees, hurricane barriers, mine tailings impoundments, or other similar water retention and/or control facilities. The Dams Sector is a vital and beneficial part of the nation's infrastructure and continuously provides a wide range of economic, environmental, and social benefits, including hydroelectric power, river navigation, water supply, wildlife habitat, waste management, flood control, and recreation.

Defense Industrial Base

The Defense Industrial Base (DIB) Sector includes Department of Defense (DoD), government, and the private sector worldwide industrial complex with the capabilities of performing research and development, design, production, delivery, and maintenance of military weapons systems, subsystems, components, or parts to meet military requirements. The DIB Sector includes tens of thousands of companies and their subcontractors who perform under contract to DoD, and companies providing incidental materials and services to DoD, as well as government-owned/contractor-operated and government-owned/government-operated facilities. DIB companies include domestic and foreign entities, with production assets located in many countries. The DIB Sector provides products and services that are essential to mobilize, deploy, and sustain military operations.

Emergency Services

The Emergency Services Sector (ESS) is a system of response and recovery elements that forms the nation's first line of defense and prevention and reduction of consequences from any terrorist attack. It is a sector of trained and tested personnel, plans, redundant systems, agreements, and pacts that provide life safety and security services across the nation via the first-responder community comprising federal, state, local, tribal, and private partners. The ESS is representative of the following first-responder disciplines: emergency management, emergency medical services, fire, hazardous material, law enforcement, bomb squads, tactical operations/special weapons assault teams, and search and rescue.

Energy

The U.S. energy infrastructure fuels the economy of the twenty-first century. Without a stable energy supply, health and welfare are threatened and the U.S. economy cannot function. More than 80 percent of the country's energy infrastructure is owned by the private sector. The energy infrastructure is divided into three interrelated segments: electricity, petroleum, and natural gas. The sector's reliance on pipelines highlights the interdependency with the Transportation Sector, and the reliance on the Energy Sector for power means that virtually all sectors have dependencies on this sector.

Government Facilities

The Government Facilities Sector includes a wide variety of buildings, owned or leased by federal, state, territorial, local or tribal governments, located domestically and overseas. Many government facilities are open to the public for business activities, commercial transactions, or recreational activities. Others not open to the public contain highly sensitive information, materials, processes, and equipment. This includes general-use office buildings and special-use military installations, embassies, courthouses, national laboratories, and structures that may house critical equipment and systems, networks, and functions.

Healthcare and Public Health

The Healthcare and Public Health Sector constitutes approximately 15 percent of the gross national product with roughly 85 percent of the sector's assets privately owned and operated. Operating in all U.S. states, territories, and tribal areas, the Healthcare and Public Health Sector plays a significant role in response and recovery across all other sectors in the event of a natural or manmade disaster.

Information Technology

The Information Technology (IT) Sector is central to the nation's security, economy, and public health and safety. Businesses, governments, academia, and private citizens are increasingly dependent upon IT Sector functions. These virtual and distributed functions produce and provide hardware, software, and IT systems and services, and—in collaboration with the Communications Sector—the Internet. The IT Sector functions are operated by a combination of entities—often owners and operators and their respective associations—that maintain and reconstitute the network, including the Internet. The Internet encompasses the global infrastructure of packet-based networks and databases that use a common set of protocols to communicate. The networks are connected by various transports, and the availability of these networks and services is the collective responsibility of the IT and Communications Sectors. The Department of Homeland Security is the sector-specific agency for the IT Sector.

National Monuments and Icons

The National Monuments and Icons (NM&I) Sector encompasses a diverse array of assets located throughout the United States and its territories. While many of these assets are listed in either the National Register of Historic Places or the List of National Historic Landmarks, all share three common characteristics: they are a monument, physical structure, object, or geographic site; they are widely recognized to represent the nation's heritage, traditions, or values, or widely recognized to represent important national cultural, religious, historical, or political significance; and their primary purpose is to memorialize or represent some significant aspect of the nation's heritage, tradition, or values, and to serve as points of interest for visitors and educational activities. NM&I Sector assets are all physical structures, objects, or geographic sites. Included as part of each asset are the operational staff and visitors that may be impacted by an attack on the asset. There are minimal cyber and telecommunications issues associated with this sector because of the nature of the assets.

Nuclear Reactors, Materials, and Waste

Nuclear power accounts for approximately 20 percent of the nation's electrical use, provided by 104 commercial nuclear reactors licensed to operate in the United States. The Nuclear Reactors, Materials, and Waste (Nuclear) Sector includes: nuclear power plants; nonpower nuclear reactors used for research, testing, and training; nuclear materials used in medical, industrial, and academic settings; nuclear fuel fabrication facilities; decommissioning reactors; and the transportation, storage, and disposal of nuclear material and waste.

Postal and Shipping

The Postal and Shipping Sector is an integral component of the U.S. economy, employing more than 1.8 million people and earning direct revenues of more than $213 billion per year. The Postal and Shipping Sector moves over 720 million messages, products, and financial transactions each day. Postal and shipping activity is differentiated from general cargo operations by its focus on small- and medium-size packages and by service from millions of senders to nearly 150 million destinations. The sector is highly concentrated, with a handful of providers holding roughly 94 percent of the market share.

Transportation Systems

The nation's transportation system quickly, safely, and securely moves people and goods through the country and overseas. The Transportation Systems Sector consists of six key subsectors, or modes: Aviation, Highway, Maritime Transportation, Mass Transit, Pipeline Systems, and Rail.

Water

Homeland Security Presidential Directive 7 (HSPD-7) designates the Environmental Protection Agency (EPA) as the federal lead for the Water Sector's critical infrastructure protection activities. All activities are carried out in consultation with the DHS and the EPA's Water Sector partners.

Methodological Approaches to Vulnerability Assessment

There are different threat approaches as part of vulnerability assessment methodologies. These threat approaches can be summarized by two main approaches: asset based and scenario based. Asset based examines the impact on individual assets (e.g., loading dock, main lobby) if attacked, whereas the scenario-based approach considers multiple potential specific sequences of events (e.g., damage or destroy building).

Vulnerability assessment methodologies are categorized as using one (or more) of the following approaches: checklist, simple rating, risk matrix, or risk equation. In some cases, a methodology may be a hybrid that incorporates elements of multiple approaches, or a range of approach options may be available to the assessor. A description of each approach category is provided next.

Checklist

Checklist-based vulnerability assessments are the simplest methodological approach. They consist of a list of questions or criteria against which the assessor compares the characteristics of the facility or asset being evaluated. The checklists may be grouped according to the various assessment elements (e.g., physical, cyber, operations security, and interdependencies). The questions may be answered yes or no or may require a qualitative response. Generally, if the answer to a question is no or if the criterion is not met, a recommended action will be requested or required. An example of a checklist methodology is the Security Vulnerability Self-Assessment Guide for Small Drinking Water Systems. Under that methodology, the water system is evaluated by answering questions for each area of concern: general security, water source protection, treatment plant protection, hazardous materials, distribution plant protection, personnel controls, information/cyber controls, and public relations information controls. Table 2.1 provides example checklist items that are included in the self-assessment guide.

Based on the understanding that a methodology is written documentation of a systematic process to be used by specialized teams or even normal employees to conduct an assessment of the risk or vulnerabilities of an asset or facility, certain

Table 2.1 Example Checklist Items from the Security Vulnerability Self-Assessment Guide for Small Drinking Water Systems

Category	Example Question
General security	Are facilities fenced, including wellhouses and pump pits, and are gates locked where appropriate?
Water sources	Are well vents and caps screened and securely attached?
Treatment plant and suppliers	Can you isolate the storage tank from the rest of the system?
Distribution	Do you control the use of hydrants and valves?
Personnel	Are your personnel issued photo-identification cards?
Information storage/ computers/ controls/maps	Is computer access "password protected" or are maps, records, and other information stored in a secure location?
Public relations	Does your water system have a procedure to deal with public information requests and to restrict distribution of sensitive information?

other forms of tools or guidance documents were not included in this survey. Guidance documents may present a general discussion of the objectives or scope of a vulnerability, risk, or security assessment; may provide typical security measures or criteria for various types of facilities or assets; or may outline potential mitigation measures for various types of vulnerabilities or assets. They do not, however, provide a step-by-step systematic, documented process to be followed in order to conduct a vulnerability assessment and/or do not contain necessary specific evaluation techniques (e.g., checklists, ranking scales, matrix categorization, or quantifiable equations). There also are assessment tools, which can be used to support vulnerability assessments, but that are not, in and of themselves, assessment methodologies. These would include software platforms and formats that can store and manipulate data or predict impact severity.

Simple Rating

Many vulnerability assessment methodologies prioritize asset vulnerabilities for potential corrective action by defining a set of measurable criteria, rating each asset (and the associated vulnerability) on each criterion, and qualitatively or quantitatively combining the individual ratings. An example of a simple, broadly applicable rating approach is a target analysis process developed and practiced by special operations forces. This process, called CARVER analysis, has been adapted and used as part of numerous vulnerability assessment methodologies. CARVER is an acronym that stands for criticality, accessibility, recoverability, vulnerability, effect, and recognizability. Each factor in the acronym typically has an associated scale (e.g., a 5-point scale), and individual assets (i.e., potential targets) are numerically rated on each factor. A rank-order of critical assets is established on the basis of the overall CARVER score (determined by summing the points assigned to the individual factors). Other "rating and weighting" schemes also are used to provide a logical and consistent basis for prioritizing vulnerabilities for importance or potential corrective actions.

Risk Matrix

A risk matrix is often used to focus vulnerability assessment results and help categorize the assets, sites, and/or systems assessed into discrete levels of risk so that appropriate protection and mitigation measures can be applied. Figure 2.2 shows a typical risk matrix, which conveys the notion that risk is a function of event severity (i.e., the severity of consequences) and the likelihood of its occurrence. Likelihood is often determined by considering the attractiveness of the targeted assets, the degree of threat, and the degree of vulnerability.

As depicted in Figure 2.2, asset vulnerabilities that have the highest likelihood of being successfully exploited (i.e., frequent) and that result in the highest severity (i.e., catastrophic), have the highest priority for vulnerability reduction actions

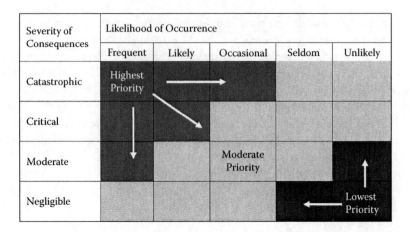

Figure 2.2   Illustrative risk matrix.

and protective measures to mitigate the risks. Similarly, asset vulnerabilities with the lowest likelihood of being exploited (i.e., unlikely) and that result in the lowest severity (i.e., negligible), have the lowest priority for mitigation. Many variations of this basic approach are used with different numbers of severity and likelihood levels, as well as definitions for those levels, to assist in focusing on the highest priority risks.

Risk Equation

Some methodologies seek a single measure that allows comparison of alternative countermeasures. Such a measure can be used to rank order or prioritize counter-measures. One approach is to calculate a risk number that is a function of prob-ability of attack, system effectiveness, and consequence. A simple formula for risk defined in this way is:

$$Risk = R = P_A \times (1 - P_E) \times C$$

where
   P_A = likelihood of occurrence (attack)
   P_E = system effectiveness [therefore, $(1 - P_E)$ = system ineffectiveness]
   C = consequence value

In some approaches, consequence value is directly addressed by assigning, for example, a low, medium, or high value. Similarly, likelihood of occurrence and system effectiveness are combined by assigning, for example, a low, medium, or high value. Finally, to calculate a "risk value," one converts low, medium, and high assignments to 0.1, 0.5, and 0.9 values, respectively, and inserts the numbers into the risk equation to calculate a risk value. Or, numerical values are determined by constructing and running models that yield likelihoods and consequences.

Another variation calls for specification of several consequences and corresponding measures (e.g., economic loss in dollars, duration of loss in hours, number of customers impacted, fatalities, and illnesses). These must then be combined in some fashion to construct a single measure of consequence value. This can be done by considering all of the measures and simply assigning a single measure that represents the set of measures (e.g., "high" for economic loss, "medium" for duration of loss, and "low" for number of customers impacted may be rated as "medium" overall).

An even more thorough approach, advocated by some, is to carefully consider the ranges that may be obtained for each of the measures and assign "weights," which are used to construct a function that yields an overall measure of the desirability of each countermeasure (or portfolio of countermeasures). This function is sometimes called a "utility function." In this approach, the measure of risk is a "utility value" and high values are preferred over low values. Therefore, the portfolio of countermeasures that yields the largest expected utility is, by definition, the most desirable (best).

Required Expertise

To carry out a vulnerability assessment, a team of experts typically needs to visit the facility to ascertain the vulnerabilities of the critical assets and the anticipated results that would be caused by their physical destruction or impairment. Depending on the objectives and scope of the assessment, VA teams may include the following types of experts (these are representative of experts used by DHS during site assistance visits to review and/or conduct vulnerability assessments):

Physical security experts who focus on the physical security of the facility, including access controls, barriers, locks and keys, badges and passes, intrusion detection devices and associated alarm reporting and display, closed-circuit television (CCTV) assessment and surveillance, communications equipment, lighting, postings, security systems wiring, and protective force personnel.

Explosive ordnance disposal (EOD) experts who examine and evaluate the vulnerability of critical assets to attacks that involve explosive devices of all kinds, including vulnerabilities to vehicle-delivered explosives and small charges.

Special operations forces or assault planning experts who focus on terrorist strategies for the most likely method of attack, including physical security vulnerabilities (e.g., fencing or CCTV gaps) and outside surveillance/positioning vulnerabilities (e.g., areas of cover for clandestine operations and positions for using long-range weapons).

Infrastructure systems experts who calculate the anticipated results of the loss of the asset as it pertains to the facility and the loss of the facility as it pertains to its specific infrastructure.

Interdependencies experts who evaluate the dependence of the facility on outside infrastructures, such as electric power, water (potable and processed) and wastewater, natural gas, steam, petroleum products, telecommunications, transportation (e.g., roads, railroads, and marine links), and banking and finance.

Operation security experts who evaluate human resources security procedures, facility engineering, facility operations, administrative support organizations, telecommunications and information technologies, publicly released information, and trash and waste handling.

Intelligence operations experts who interact with local law enforcement and intelligence/security personnel to determine if there are potential terrorist or criminal elements in the region that may have an interest in the facility.

Outline of Risk Management Steps

This section presents an outline of the risk management process that has been applied by the Department of Energy, Department of Homeland Security, and private sector. Table 2.2 provides an overview of representative steps in a comprehensive, asset-based vulnerability assessment methodology. This includes countermeasure (actions taken to reduce or eliminate vulnerabilities) and risk management considerations. The methodologies included in this survey address, to a greater or lesser degree, some or all of these steps.

The following sections describe the steps of the risk management process in more detail. Where appropriate, the steps contain checklists of questions that could be used to guide the implementation of a risk management program.

Step 1: Identify Critical Assets and the Impacts of Their Loss

Estimates of the potential consequences, including economic implications, of not mitigating identified vulnerabilities or addressing security concerns are necessary to effectively apply risk management approaches to evaluate mitigation option and security recommendations. Outages because of security failures could degrade an energy facility's reputation and place the community served at risk to economic losses or even losses of property and life.

In addition, the modern energy facility's telecommunication and computer network has many external connections to public and private networks. Such connections are used to communicate with customers and offer new electronic services, such as online billing and payment. Cyber security should be a primary concern, especially for utilities that operate in this interconnected environment. An IT security architecture may need to be developed.

Possible critical assets include people, equipment, material, information, installations, and activities that have a positive value to an organization or facility. People include energy facility executives and managers, security personnel, contractors and vendors, and field personnel. Equipment includes vehicles and other transportation

Table 2.2   General Vulnerability Assessment Process

Step	Description	Considerations
1	Identify critical assets and the impacts of their loss	• Identify the critical functions of the facility. • Determine which assets perform or support the critical functions. • Evaluate the consequences or impacts to the critical functions of the facility from the disruption or loss of each of these critical assets.
2	Identify what protects and supports the critical assets	• Identify the components of the physical security system (e.g., perimeter barriers, building barriers, intrusion detection, access controls, and security forces) that protect each asset. • Identify the critical internal and external infrastructures (e.g., electric power, petroleum fuels, natural gas, telecommunications, transportation, water, emergency services, computer systems, air handling systems, fire systems, and SCADA systems) that support the critical operations of each asset (interdependencies). • Identify sensitive information about the facility and its operation that must be protected.
3	Identify and characterize the threat	• Gather threat information and identify threat categories and potential adversaries. • Identify the types of threat-related undesirable events or incidents that might be initiated by each threat or adversary. • Estimate the frequency or likelihood of each threat-related undesirable event or incident based on historical information. • Estimate the degree of threat to each critical asset for each threat-related undesirable event or incident.
4	Identify and analyze vulnerabilities	• Identify the potential vulnerabilities of each asset to each threat or adversary.

Continued

Table 2.2 (*Continued*)   General Vulnerability Assessment Process

Step	Description	Considerations
4 cont.	Identify and analyze vulnerabilities	• Identify the existing measures intended to protect the critical assets and estimate their levels of effectiveness in reducing the vulnerabilities of each asset to each threat or adversary. (Step 2 provides a starting point for this activity.) • Estimate the degree of vulnerability of each critical asset for each threat-related undesirable event or incident and thus each threat or adversary.
5	Assess risk and determine priorities for asset protection	• Estimate the effect on each critical asset from each threat or adversary taking into account existing protective measures and their levels of effectiveness. • Determine the relative degree of risk to the facility in terms of the expected effect on each critical asset (a function of the consequences or impacts to the critical functions of the facility from the disruption or loss of the critical asset, as evaluated in Step 1) and the likelihood of a successful attack (a function of the threat or adversary, as evaluated in Step 3, and the degree of vulnerability of the asset, as evaluated in Step 4). • Prioritize the risks based on the relative degrees of risk and the likelihoods of successful attacks using an integrated assessment.
6	Identify mitigation options, costs, and trade-offs	• Identify potential mitigation options to further reduce the vulnerabilities and thus the risks. • Identify the capabilities and effectiveness of these mitigation options. • Identify the costs of the mitigation options. • Conduct cost-benefit and trade-off analyses for the various options. • Prioritize the alternatives for implementing the various options and prepare recommendations for decision makers.

equipment, maintenance equipment, operational equipment, security equipment, and IT equipment (computers and servers). Material includes tools, spare parts, and specialized supplies. Information includes employee records; security plans; asset lists; intellectual property; patents; engineering drawings and specifications; system capabilities and vulnerabilities; financial data; and operating, emergency, and contingency procedures. In addition to the operational installations that make up the energy infrastructure itself; installations include headquarters offices; field offices; training centers; contractor installations; and testing, research, and development laboratories. Activities include movement of personnel and property, training programs, communications and networking, negotiations, and technology research and development.

The energy facility, the local government, and energy industry associations have roles and responsibilities for identifying assets, effects of asset loss, vulnerabilities, threats, and risk mitigation options. Coordination among energy facilities; local, state, and federal agencies; and energy industry associations is crucial to this process.

Energy facilities need to identify the critical functions of the facility, and determine which physical and cyber assets perform or support the critical functions. The key assets identified should be related to the criticality of overall operations of the individual facilities. Potential assets include substations, transmission lines, pipelines, critical valve nests, power plants, pump stations, city gate stations, compressor stations, storage installations, interconnections, energy control centers, energy management systems (EMS), SCADA systems, remote monitoring and control units (remote terminal units [RTU]), communications systems linking RTUs and energy control centers, certain backup systems, and e-commerce capabilities. They should evaluate the consequences or impacts to the critical functions of the energy facility from the disruption or loss of each of these critical assets and prioritize the critical assets based on these.

Not all assets and activities warrant the same level of protection. The cost of reducing risk to an asset must be reasonable in relation to its overall value. The value, however, does not need to be expressed in dollars. A potential loss can be stated in terms of human lives or the impact on the local or state economy.

The first set of questions is designed to guide the process of identifying the critical functions of the energy facility and the assets that perform or support them, and evaluating the potential consequences of disruptions or loss of these critical assets.

Criticality Criteria (Functions and Assets)
- What critical mission activities take place at the energy facility or its remote sites?
- What critical or valuable equipment is present at the facility or its remote sites?
- Where are the critical assets located?
- Have people, installations, and operations been considered when assessing assets?

- Have cyber networks and system architectures (e.g., SCADA systems, business e-mail, and e-commerce) been documented fully?

Criticality Criteria (Impacts of Loss)

- What would the energy facility lose if an adversary obtained control of a specific asset?
- What affects would be expected if a specific asset were compromised?
- Is the asset still valuable to the energy facility once an adversary has it?
- What is the potential for immediate and significant local impacts due to the loss of the asset?
- What is the potential for loss of energy supply to civilian areas?
- What facility personnel, tenants, customers, and visitors could be affected by the loss of the asset?
- What would be the impact on people's lives and on national or local security due to the loss of the asset?
- What would be the financial impacts to the energy facility and the local community?

Criticality Criteria (Asset Value)

- Is there little or no redundant capacity or capability to mitigate the loss of the asset?
- What is the potential for cascading effects (e.g., to other interdependent infrastructures or industries) due to the loss of the asset?
- Do any special situations need to be considered regarding the loss of the asset, such as the status of hospitals, life support systems, or emergency services that depend on the energy infrastructure supported by the asset?
- What is the potential for catastrophic effects (weapons of mass destruction levels impact)?
- What did it cost to develop the asset?
- Would the energy facility need an extended period to make repairs to the asset?
- How does the need for protecting the asset compare with other assets also considered critical?

Once the assets critical to the operation of the energy facility have been identified and characterized, an impact assessment must be carried out to describe the consequences of losses if an undesirable event occurs. The degree of impact should be quantified by using a relative impact or criticality rating criteria and a consistent rating scale. (An example of a scale for rating criteria is presented in Step 5.) The assets are then ranked in terms of criticality.

Step 2: Identify What Protects and Supports the Critical Assets

The existing protection of critical assets provided by the physical security system and the dependence of the critical assets on both external and internal infrastructures must be known to evaluate the vulnerabilities of the assets to threats or adversaries. In addition, operating procedures and other sensitive information, which if available to adversaries might jeopardize critical assets, must be identified as their availability can also affect the vulnerabilities of assets.

Physical Security Systems

Physical security systems are used to protect energy facilities and their assets from unauthorized individuals and outside attacks. Such systems usually include perimeter barriers, building barriers, intrusion detection, access controls, and security forces.

Infrastructure Interdependencies

Today's energy facilities depend on many different infrastructures to support their critical functions and assets. These infrastructure interdependencies must be identified and the adequacy of security measures that are in place to protect and back up these infrastructures must be evaluated. Typically, these supporting infrastructures include:

- Electric power supply and distribution
- Petroleum fuels supply and storage
- Natural gas supply
- Telecommunications
- Transportation (road, rail, air, and water)
- Water and wastewater
- Emergency services (fire, police, and emergency medical)
- Computers and servers
- Heating, ventilation, and air conditioning (HVAC) systems
- Fire suppression and firefighting systems
- SCADA systems

The electric power supply and distribution infrastructure can include the local electrical distribution utility, facility-operated electric generation equipment, backup generators fueled by natural gas or petroleum fuels, uninterruptible power supplies (UPSs), and the associated switching and distribution hardware. The petroleum fuels supply and storage infrastructure includes on-site storage as well as local suppliers, storage terminals, and the entire petroleum industry. The telecommunications infrastructure includes commercial telephone, fiber optic, and satellite networks and facility-owned radio, telephone, microwave, and fiber-optic pathways. Computers and servers, HVAC systems, fire suppression and firefighting systems, and SCADA systems tend to be operated by the energy facility and, in turn, depend on the other infrastructures such as telecommunication, electric

44 ■ *Principles of Emergency Management and Operations*

power supply and distribution, petroleum fuels supply and storage, natural gas supply, water and wastewater, and emergency services.

Sensitive Information

Protecting operating procedures and other sensitive information, the release of which might jeopardize an energy facility and its assets, is the objective of OPSEC programs. OPSEC programs utilize tools such as employee background checks, trash handling procedures, telephone policies, and IT (computer) security to protect against both industrial espionage and deliberate disruption of critical assets and functions.

The second set of questions is designed to guide the process of identifying the existing components of the physical security system that protect the critical assets, the critical infrastructure systems that support the critical assets, and the operating procedures and sensitive information that must be protected to avoid jeopardizing the critical assets.

Energy Facility and Critical Asset Protection (Physical Assets)
- What department or person has overall responsibility for security or is that responsibility spread over many departments or people with shared responsibilities for security along with their other responsibilities?
- What perimeter barriers (e.g., fences, gates, vehicle barriers) protect the energy facility as a whole and the individual critical assets, and what levels of protection do they provide?
- What building barriers (e.g., walls, roof/ceiling, windows, doors, locks) protect each critical asset and what levels of protection do they provide?
- What is the status of the intrusion detection that protects each critical asset (e.g., intrusion sensors, alarm deployment, alarm assessment, alarm maintenance) and what level of protection does it provide?
- What is the status of the access control that is used at each critical asset (e.g., personnel access, vehicle access, contraband detection, access point illumination) and what level of protection does it provide?
- What is the nature of the security force (both the protective force and appropriate local law enforcement agencies) that protects each critical asset (e.g., number, training, armament, communications) and what level of protection does it provide?
- What types of undesirable events (e.g., surreptitious forced entry, technical implant, theft of sensitive information or materials) are protected against?
- During which hours of the day and under what conditions are the various components of the physical security system effective?
- Over what areas do the various components of the physical security system provide protection?

- What is the history of reported malfunctions of the various components of the physical security system?
- What is the correlation of the effectiveness of the various components of the physical security system to security incident reports that may indicate that the system was defeated?
- Have liaisons and working relationships been established with the local government and its departments, such as police, fire, emergency medical services, and public works?

Energy Facility and Critical Asset Protection (Infrastructure Interdependencies)

- Which infrastructures (both internal and external) are essential for a specific critical asset to be able to carry out its critical functions?
- What external utility or internal department and equipment is the normal provider of each essential infrastructure for each critical asset and how is each infrastructure connected to each asset (e.g., the types and pathways of power lines, pipelines, and cables)?
- What alternatives (e.g., redundant systems, alternative suppliers, backup systems established work-around plans) are available if the normal providers of the essential infrastructures are disrupted, and how long can the alternatives support the critical functions of the assets?
- What is the potential for interdependency effects on external infrastructures (i.e., effects on the energy, telecommunications, transportation, water and wastewater, banking and finance, emergency services, and government services infrastructures)?

Energy Facility and Critical Asset Protection (Sensitive Information)

- What types of information about the energy facility, its assets, and its operations should be considered critical or sensitive information?
- What are the methods and means by which sensitive information might fall in the wrong hands, such as via disgruntled employees; access to the facility by the public; outside construction, repair, and maintenance contractors; press contacts; briefings and presentations; public testimony; Internet information; paper and material waste; telecommunications system taps; and cyber (computer) intrusions?
- What are the existing policies and procedures that are used to protect sensitive information, such as employee background checks, disciplinary procedures, security training, trash handling procedures, paper waste handling procedures, salvage material handling procedures, dumpster control, telephone policies, and IT (computer) security.

Once energy facilities have identified their existing physical protective measures, they should coordinate with their respective local governments and law enforcement agencies to ensure that the level of protection and response that they expect will be forthcoming. They should also coordinate with the critical external infrastructure providers to ensure that the robustness and redundancy that they depend on will continue to be provided. The objective of these coordination efforts is to ensure that roles for response and recovery from a disruption are understood by all so that quick and effective measures can be taken when problems occur.

In addition, local and state governments can assist energy facilities in infrastructure restoration activities. Potential support can come in many areas, such as maintaining critical spare parts, assisting with special equipment, working with the emergency telecommunications spectrum, securing easy access to the site of the disruption for repair crews and needed equipment, working out mutual assistance programs with other energy providers, and supplying temporary staffing.

The energy facility should also check with local and state governments to ensure that critical information about their facility, its assets, and its operations will not be released to the general public in any future additions to public Internet sites, press releases, or public hearings.

Step 3: Identify and Characterize the Threat

In order to put the information about the critical assets of the energy facility to use in a quantitative risk assessment, the potential threats and adversaries that may be expected must be identified and quantified. The set of questions provided in this section serves as guidance for evaluating the threat environment to which the energy facility could be exposed and establishing qualitative or quantitative threat ratings for each critical asset. The goals of the threat assessment are to understand, from the adversary's point of view, the adversary's capabilities and intent to collect critical information.

The federal agencies (e.g., Department of Energy [DOE] and the Federal Bureau of Investigation [FBI]), state governments, and energy industry associations each collect threat information. This information should be shared among these groups and with the local energy facilities in order to have the most comprehensive and updated threat information possible. In addition, threats to energy facilities could affect state and local assets. State and local governments have access to law enforcement and intelligence data. This information should be integrated and shared, together with any information that the energy industry associations and energy facilities collect.

This third set of questions is to be used to identify and evaluate the threat environment to which an energy facility may be exposed.

Intent and Capabilities of Adversaries
- What types of adversaries are expected?
- Who are the specific adversaries expected?
- What are the specific goals and objectives of each adversary?
- Which are the critical assets that each specific adversary is aware?
- Does each specific adversary know enough about the asset to plan an attack?

- What are the possible modes of attack (e.g., explosives or incendiary devices delivered by car, truck, boat, rail, mail, individuals, or standoff weapons; aircraft impacts; sabotage of equipment or operations; assaults by a lightly or heavily armed individual attacker or team of attackers; theft, alteration, or release of information, materials, or equipment; contamination by chemical agents, biological agents, or radioactive material; and cyber attacks) each adversary might use?
- Are there other, less risky means for a specific adversary to attain his or her goals?
- What is the probability that an adversary will choose one method of attack over another?
- What specific events might provoke a specific adversary to act?

Information concerning potential threats and adversaries can be gathered about potential threats and adversaries by:

■ Joining a threat analysis working group that includes local, county, state, and federal agencies, the military, and other industry partners
■ Obtaining access to the National Infrastructure Protection Center (NIPC), Analytical Services, Inc. (ANSER), FBI-sponsored InfraGuard, Carnegie Mellon University's CERT®, or other information system security warning notices
■ Initiating processes to obtain real-time information from the field (e.g., on-duty offices, civilian neighborhood watch programs, local businesses, other working groups in the area)
■ Arranging for threat briefings by local, state, and federal agencies
■ Performing trend analyses of historical security events (both planned and actual)
■ Creating possible threat scenarios based on input from the threat analysis working group and conducting related security exercises

Step 4: Identify and Analyze Vulnerabilities

In addition to identifying the critical assets of the energy facility, the impact of their disruption, the present protection provided, and the potential threats against them, the vulnerability of those assets to the potential threats must be quantified, at least to some extent, to determine the overall risk to the assets.

There are various types of vulnerabilities, such as physical, technical/cyber, and operational. An energy facility, including perimeter barriers (fences, walls, gates, landscape, sewers, tunnels, parking areas, alarms), compound area surveillance (closed-circuit television, motion detectors, lighting), building perimeters (walls, roofs, windows, doors, shipping docks, locks, shielded enclosures, access control, alarms), and building interiors (doors, locks, safes, vents, intrusion sensors, motion sensors), is subject to physical vulnerabilities. Electronic equipment, such as acoustic equipment, secure telephones, computers and computer networks, and radio-frequency equipment, are subject to technical or cyber vulnerabilities. The guard force, personnel procedures, and operational procedures are subject to operational vulnerabilities.

Various characteristics of assets, including any existing protection identified in Step 2, may affect their susceptibility to attacks and must be considered when identifying susceptibilities. Such asset characteristics include building design; equipment properties; personal behavior; locations of people, equipment, and buildings; and operational and personnel practices.

Both energy facilities and local governments should be concerned with identifying and analyzing vulnerabilities. Energy facilities should analyze the vulnerabilities of their physical and cyber systems. Local governments should coordinate management of the vulnerabilities of the energy infrastructure, including individual energy facilities, that support government and community operations and assets.

This fourth set of questions is to be used to evaluate the vulnerability of the critical energy infrastructure assets to the potential threats and to establish qualitative or quantitative vulnerability ratings for each asset.

Energy Facility and Critical Asset Vulnerabilities
- How susceptible is each critical asset to physical attack if readily available weapons (guns, normal ammunition, vehicle, simple explosives) were used?
- How susceptible is each critical asset to physical attack if difficult-to-acquire weapons (assault rifles, explosive ammunition, rocket launchers, biological or chemical agents, aircraft, sophisticated explosives) were used?
- How susceptible is each critical asset to physical attack from insiders?
- Are any of the critical assets unprotected? If so, describe them.
- Are any of the critical assets minimally protected? If so, describe them.
- How susceptible is each critical asset to cyber attack?

Step 5: Assess Risk and Determine Priorities for Asset Protection

Scales for the rating criteria identified in the first four steps (asset criticality in terms of the impact of loss or disruption, threat characteristics, and asset vulnerability) must be developed. The concept of criteria development is presented below in the form of a generic example. Those that conduct an actual assessment should define rating scales that are appropriate to the specific assessment.

Using the individual rating values assigned to each combination of asset criticality, threat, and vulnerability, a relative degree of risk or a risk rating can be established for each asset for one or more postulated adverse events or consequences that could result from an attack by the identified adversary. Often a multiplicative approach involving the three rating criteria is used to obtain a risk rating:

Risk Rating = Impact Rating × Threat Rating × Vulnerability Rating

An additional scale must be developed to assign a qualitative overall risk level from the quantitative risk rating. The risk ratings or risk levels are used to prioritize the assets for the selection and implementation of security improvements to achieve an acceptable overall level of risk at an acceptable cost.

The following should be considered when developing and using rating criteria:

- Subject-matter expert opinions and perspectives should be documented. The team involved in the assessment should reflect a variety of different perspectives, and the team should work toward reaching a consensus regarding a set of priorities.
- Information should be presented in a usable format (e.g., table, matrix, or spreadsheet).
- Assumptions should be documented.

A generic example of possible scales for the rating criteria is presented next in the form of a set of tables to illustrate the concept. These or similar tables are used to establish qualitative or quantitative criticality, threat, and vulnerability ratings for each critical asset.

Asset Impact/Criticality Rating Criteria

Each critical asset that is identified in Step 1 of the risk management process is assigned an impact rating value that reflects the importance or criticality of a loss or disruption of that asset with regard to the continued operation of the energy facility or other organization being assessed. In Table 2.3, a quantitative criticality rating scale of 0 percent to 100 percent is used, which corresponds to qualitative criticality levels of critical, high, medium, and low.

Threat Rating Criteria

The individual potential threats against the assets of the energy facility or other organization being assessed that are identified in Step 3 are assigned a threat rating value that reflects the magnitude of the threat. In Table 2.4, a quantitative threat rating scale of 0 percent to 100 percent is used, which corresponds to qualitative threat levels of critical, high, medium, and low.

Vulnerability Rating Criteria

The vulnerabilities of the assets in terms of in-place measures to protect those assets that are identified in Step 2 are assigned a vulnerability rating value that reflects the extent to which the asset is protected against each threat identified in Step 3. In Table 2.5, a quantitative vulnerability rating scale of 0 percent to 100 percent is

Table 2.3 Asset Impact/Criticality Rating Criteria

Criticality Level	Description	Rating Scale (%)
Critical	Indicates that compromise of the asset would have grave consequences leading to loss of life or serious injury to people and disruption of the operation of the energy facility. It is also possible to assign a monetary value or some other measure of criticality.	75–100
High	Indicates that compromise of the asset would have serious consequences that could impair continued operation of the energy facility.	50–75
Medium	Indicates that compromise of the asset would have moderate consequences that would impair operation of the energy facility for a limited period.	25–50
Low	Indicates little or no impact on human life or the continuation of the operation of the energy facility.	1–25

Table 2.4 Threat Rating Criteria

Threat Level	Description	Rating Scale (%)
Critical	Indicates that a definite threat exists against the asset and that the adversary has both the capability and intent to launch an attack, and that the subject or similar assets are targeted on a frequently recurring basis.	75–100
High	Indicates that a credible threat exists against the asset based on knowledge of the adversary's capability and intent to attack the asset and based on related incidents having taken place at similar assets or in similar situations.	50–75
Medium	Indicates that there is a possible threat to the asset based on the adversary's desire to compromise the asset and the possibility that the adversary could obtain the capability through a third party who has demonstrated the capability in related incidents.	25–50
Low	Indicates little or no credible evidence of capability or intent and no history of actual or planned threats against the asset.	1–25

Table 2.5   Vulnerability Rating Criteria

Vulnerability Level	Description	Rating Scale (%)
Critical	Indicates that there are no effective protective measures currently in place and adversaries would be capable of exploiting the critical asset.	75–100
High	Indicates that although there are some protective measures in place, there are still multiple weaknesses through which adversaries would be capable of exploiting the asset.	50–75
Medium	Indicates that there are effective protective measures in place; however, one weakness does exist that adversaries would be capable of exploiting.	25–50
Low	Indicates that multiple layers of effective protective measures exist and essentially no adversary would be capable of exploiting the asset.	1–25

used, which corresponds to qualitative vulnerability levels of critical, high, medium, and low.

Step 6: Identify Mitigation Options, Costs, and Trade-Offs

The ultimate goal of a risk management process is to select and implement security improvements to achieve an acceptable overall risk at an acceptable cost. Step 5 of the risk management process prioritizes the combinations of assets and threats by the risk ratings or risk levels. This, in turn, helps to identify where protective measures against risk are most needed.

In this step, potential measures to protect critical assets from recognized threats are identified, specific programs to ensure that appropriate protective measures are put into place are established, and appropriate agencies and mechanisms needed to put protective measures in place are identified. Protective measures that can address more than one threat or undesirable event should be given special attention.

A variety of approaches to developing protective measures exists. Protective measures can reduce the likelihood of a failure due to an attack by adding physical security. Protective measures can also be implemented to prevent or limit the consequences of a failure or to speed the recovery following a failure, no matter what the cause of that failure.

Best practices and lessons learned from DOE's Vulnerability Survey and Analysis Program provide some general actions, activities, and recommendations that can help identify appropriate potential mitigations measures. Some of these are listed next.

■ The trend in IT until very recently has been to outsource more and more functions. Since the events of September 11, 2001, outsourcing is becoming less popular again. If possible, cyber security should remain as an enterprise function and should not become a contractor function.

■ Logging and reporting should be enabled on IT network routers and firewalls to gain a better understanding of user access and interactions with remote systems.

■ Sensitive and confidential documents should not be placed on Web sites. Appropriate document review, classification, and access controls should be implemented. This practice should apply to documents and other information that is found in newsgroups, media sites, and other linked sites.

■ Security measures, such as traffic filtering, authorized controls, encryption and access controls, minimizing or disabling of unnecessary services and commands, minimizing banner information, and e-mail filtering and virus control, should be implemented.

■ A formal process for accessing relevant threat information and for contacting the proper government and law enforcement agencies should be instituted (if it does not already exist), and reviewed and updated on a regular basis. The energy facility may need to work with government to obtain security clearances for appropriate personnel.

■ Appropriate security measures (e.g., access controls, barriers, badges, intrusion detection devices, alarm reporting and display, closed-circuit television cameras, communication equipment, lighting, and security officers) should be implemented.

■ Top management support is critical in ensuring a successful security program.

■ Security training programs should be formalized.

■ Procedures for escorting contractors and visitors into sensitive areas should be enhanced and enforced.

■ Security should be incorporated in the company goals as well as in its corporate culture.

■ The foundation for security is well-informed employees acting responsibly.

■ A formal review process should be established for all information released to the public, particularly through the energy facility's Web site. A periodic review of "public" information should be performed to audit the effectiveness of information protection policies.

■ The energy facility should be careful about disseminating sensitive information to the press or competitors. Only minimal information should be made available about personnel (especially executives).

■ Security training and awareness should be provided to all employees on a regular basis.
■ At a minimum, an annual audit of overall security should be conducted.

Some illustrations specific to energy facilities, including large utilities not specifically considered in this report, are listed next as an example of specific protective measures that can be implemented. The examples are grouped by type.

Measures to Prevent Damage
   – Harden key installations and equipment—protect critical equipment with walls or below-grade installations, physically separate key pieces of equipment, and toughen the equipment itself to resist damage.
   – Install surveillance systems (e.g., video cameras, motion detectors) around key installations that are monitored and coupled with rapid-response forces.
   – Maintain security guards at key installations.
   – Improve communication with law enforcement agencies, especially local law enforcement agencies and the local FBI office, to obtain threat information and coordinate responses to emergencies.
Measures to Limit Consequences
   – Improve emergency plans and procedures for continued operation during undesirable events and ensure that operators are trained to implement these contingency plans.
   – Modify the physical system—improve control centers and protective devices, increase redundancy of key equipment, and increase reserve margins.
Measures to Speed Recovery
   – Conduct contingency planning for restoration of service, including identification of potential spare parts and resolution of legal uncertainties.
   – Clarify the legal and institutional framework for sharing reserve equipment.
   – Stockpile critical equipment (e.g., transformers, pumps, compressors, regulators) or any specialized materials (e.g., cables, pipe sections) needed to manufacture critical equipment or make repairs.
   – Assure availability of adequate transportation for stockpiles of very heavy equipment by maintaining a database of rail and barge equipment and adapting Schnabel railcars to fit all needed types of large pieces of equipment, if necessary.
   – Monitor domestic manufacturing capability to assure adequate repair and manufacture of key equipment in times of emergency.
   – Investigate mutual aid agreements with vendors, industry associations, or large nearby energy companies.
   – Establish backup arrangements with contractors for emergency services and other emergency support.

General Mitigation Measures to Reduce Vulnerability
- Emphasize inherently less vulnerable technologies and designs when practical, such as using standardized equipment.
- Move toward an inherently less vulnerable bulk energy system (e.g., smaller generators near loads, local storage) as new installations are planned and constructed.

As indicated earlier, local governments should coordinate energy facility activities related to risk management. The following questions can help guide local and state governments through the risk management process.

Local and State Government's Role in the Risk Management Process
- Has the local or state government identified any critical issues or vulnerabilities regarding its energy infrastructure?
- If the local or state government has identified critical issues, what are they and why are they critical?
- Has the local or state government developed plans to counter these vulnerabilities?
- Has the local or state government coordinated information with other local energy facilities, local law enforcement agencies, and others concerning these vulnerabilities?

Conclusion

Vulnerability assessment methodologies generally evaluate vulnerabilities by broadly considering the threat, existing protective measures, and consequences that result if an asset is attacked (asset-based approach). Alternatively, multiple potential sequences of attack events can be considered to evaluate the likelihood that the current protective measures at a facility will be able to successfully deter, detect, and/or delay an attack (scenario-based approach).

Protecting and ensuring the continuity of the critical infrastructure and key resources (CIKR) of the United States are essential to the nation's security, public health and safety, economic vitality, and way of life.

References

Critical Infrastructure and Key Resources. www.dhs.gov/files/programs/gc_1189168948944.shtm.
Energy Infrastructure Risk Management Checklists for Small and Medium Sized Energy Facilities, U.S. Department of Energy, Office of Energy Assurance, August 2002.

Survey of Vulnerability Assessment Methodologies, U.S. DHS Protective Security Division, October 2003.

The White House, Homeland Security Presidential Directive/HSPD-7: Critical Infrastructure Identification, Prioritization, and Protection, Washington, DC, December 17, 2003. www.fas.org/irp/offdocs/nspd/hspd-7.html

Chapter 3

Developing a Planning Team

Michael Fagel

Contents

The Emergency Planning Team

Emergency planning should be a team effort because disaster response nearly always requires coordination among every department within an organization. Planning as a team helps to ensure that everyone having the expertise to contribute in a response is a part of the process. Team planning also helps to ensure the buy-in of all key players or stakeholders.

Who Should Be on the Planning Team?

Different types of emergencies require different expertise and response capabilities. The specific individuals and organizations involved in a response will vary by the type of emergency or disaster. Some emergencies may involve only an internal response, whereas others may require the response of external agencies, such as law enforcement, fire services, and emergency medical services (EMS). These external agencies should be involved in large-scale planning, as well as other agencies, such as those responsible for hazardous materials (hazmat) response, should be included as well, as they may be needed in a response to, for example, an emergency involving a refrigeration leak or a different hazardous material.

Table 3.1 presents a breakdown of potential team members. Not all organizations will have such a breakdown of personnel, and dependent on the size of the organization, one person may be responsible for multiple roles. All individuals do not need to be included in every aspect of the planning process, but they should be consulted in areas that affect them directly or for which they will be responsible.

Getting the Team Together

Getting all of the stakeholders in the planning process together and to take an active interest in the planning process will be an arduous task. To schedule meetings with so many participants may be even more difficult. It is critical, however, to have everyone's participation in the planning process, at least in the early stages and at critical points along the way, to draw from their expertise and ensure their buy-in to the plan.

The expertise and knowledge that participants bring of their departments' needs and resources are crucial to developing an accurate plan that considers the entire organization's needs and the resources that could be made available in an emergency. It is definitely to the organization's benefit to have the active participation of all key players.

There are several steps that can be taken to gain cooperation from participants and to gain their involvement in the planning process.

Plan ahead—Give the planning team plenty of notice of where and when the planning meeting will be held. If time permits, survey the team members to find the time and place that will work for them.

Provide information about team expectations—Explain why participating on the planning team is important to the participants' departments and to the organization as a whole. Show the team members how they will contribute to a more effective emergency response.

Involve the chief executive officer—Ask the CEO to sign the meeting notice. This will send a clear message that emergency planning is important to the organization and that the participants are expected to attend.

Table 3.1 Possible Emergency Planning Team Members

Individuals/Organizations	What They Bring to the Planning Team
Chief executive officer (CEO) or designee	Support for the emergency planning process, policy guidance and decision-making capability, and the authority to commit the organization's resources
Plant manager or designee	Knowledge of the Incident Command System (ICS), knowledge of organizational procedures, safety requirements, hazardous materials response requirements; knowledge of external agencies or the abilbity to interface with them
Safety director	Knowledge of the organization's safety planning and steps already taken, as well as the ability to interact with all departments
Departmental supervisors or designee	Knowledge of individual department procedures, safety requirements, and personnel
EMS/Employee health director or designee	Knowledge of medical treatment requirements for a variety of situations, knowledge of medical treatment facility capabilities, and specialized personnel and equipment resources
Hazardous materials coordinator	Knowledge of hazardous materials that are produced, stored, or transported into or off the organization's campus, knowledge of Environmental Protection Agency (EPA), Occupational Safety and Health Administration (OSHA), and Department of Transportation (DOT) requirements for producing, storing, and transporting hazardous materials, knowledge of how to respond to hazardous materials incidents
Mutual aid partners	Specialized personnel and equipment resources, and additional personnel and equipment resources; includes fire, EMS, and law enforcement for the community in which the facility is located as well as other community departments as needed

Allow flexibility in scheduling after the first meeting—Not all team members will need to attend all meetings. Task forces and subcommittees can complete some of the work. Where this is the case, gain concurrence on timeframes and milestones, but let the subcommittee members determine when it is most convenient to meet.

It also may be beneficial to talk to emergency managers from similar organizations in the community to gather input on how to gain and maintain interest in the planning process, as well as to see what practices they already have in place that work or have already been tested.

Team Operation

Unlike working alone, working with personnel from other departments to plan for emergencies requires some give and take—in other words, collaboration. Collaboration is the process in which people work together as a team on a common mission—in this case, the development of a company emergency operations plan (EOP). Successful collaboration requires:

- A commitment to participate in shared decision making
- A willingness to share information, resources, and tasks
- A professional sense of respect for individual team members

Although collaboration among EOP planning team members may be difficult, it benefits the organization by strengthening the overall response to disaster. Collaboration can:

- Expand resource availability through sharing
- Enhance problem solving through the cross-pollination of ideas

Stages of Team Formation

Collaboration does not come automatically. Building a team that works well together takes time and effort and typically evolves through five stages:

1. Forming—Individuals come together as a team. During this stage, the team members may be unfamiliar with one another and uncertain of their roles on the team.
2. Storming—Team members may become impatient, disillusioned, and may disagree.

3. Norming—Team members accept their roles and make progress toward the goal.
4. Performing—Team members work well together and make progress toward the goal.
5. Adjourning—Their task accomplished, team members may feel pride in their achievement and some sadness that the experience is ending. It is important to remember that the team should remain intact for the purpose of exercising and revising the plan.

Team Roles

To keep the team focused throughout the planning stages, it is important for team members to assume roles. Perhaps the most important role is that of team leader. The team leader initiates appropriate team-building activities that move the team through each stage and toward its goal. Other team roles may include:

■ The task master, who identifies the work to be done and motivates the team
■ The innovator, who generates original ways to get the group's work done
■ The organizer, who helps the group develop plans for getting the work done
■ The evaluator, who analyzes ideas, suggestions, and plans made by the group
■ The finisher, who follows through on plans developed by the team

Together, all members contribute toward making the team productive.

Characteristics of an Effective Team

You will know that the planning team is on track when it displays the following characteristics:

■ Works toward a common goal
■ Accepts the leader who provides direction and guidance
■ Communicates openly
■ Resolves conflicts constructively
■ Displays mutual trust
■ Shows respect for each individual and his or her contributions

Summary

Emergency planning requires collaboration from a variety of individuals and organizations. The benefits of collaboration far outweigh the difficulties that you will face during the planning process. Some of the benefits include elimination of duplication of services, expanded resource availability, and enhanced problem solving.

The planning process can be made easier by planning ahead, providing information about team expectations, and allowing flexibility in scheduling. It may be to your benefit to talk to emergency managers from other communities to gain their input on the planning process.

Successful team operation requires a commitment to participate in shared decision making, a willingness to share information, and a professional sense of respect for individual team members.

Team formation usually takes place through five stages: forming, storming, norming, performing, and adjourning.

You will know that you have an effective planning team when members agree on and work toward a common goal; provide open communication; and display constructive conflict resolution, mutual trust, and respect for other team members.

Chapter 4

Responder Health and Safety

Michael Steinle

Contents

Introduction

As discussed in previous chapters, the former National Response Plan (NRP), the National Response Framework, and the National Incident Management System (NIMS) provide fundamental guidance for preparedness and response. Health and safety provisions for emergency response should be devised within the framework of these guidance documents and should be consistent with the Incident Command System (ICS). The incident commander and safety officer must work to ensure that responders perform tasks only after they thoroughly assess the associated hazards and implement effective hazard control measures. The information presented in this chapter is intended to provide guidance to people who may serve in the incident commander or safety officer roles during emergency response operations.

The Occupational Safety and Health Administration (OSHA); the National Fire Protection Association (NFPA); and many other government agencies, not-for-profit organizations, and trade associations have set standards for protecting people in the workplace. Traditional health and safety doctrines have been premised on three types of controls: (1) administrative controls, (2) engineering controls, and (3) use of personal protective equipment (PPE). Moreover, health and safety doctrines have been largely premised on planned workplace hazards. In other words, devising protective measures in traditional workplaces is done through job hazard analysis on tasks that are relatively static. The task conditions are known, they do not change drastically from day to day, and employees perform the same tasks and generally understand them implicitly.

However, conditions during emergency response operations can differ drastically from the aforementioned conditions. In emergencies, the hazards are likely not fully understood, particularly in initial response operations, may change drastically throughout response operations, and are less likely to be mitigated through the standard administrative or engineering controls. Responders may also be less familiar with hazards associated with a particular response scenario due to infrequent exposure to such hazards. The purpose of this chapter is to adapt relevant health and safety doctrines to emergency response operations, both programmatically and systematically.

This chapter highlights three significant characteristics of an effective program to provide protective measures:

1. Systemic processes for health and safety planning and communicating precautionary actions to limit exposure to hazards
2. Programs for personal protective equipment (PPE) and other mitigation strategies
3. Medical protective measures, including medical monitoring and supplying prepositioned medication for first responders and their immediate families

Defining Essential Personnel

Who essential personnel will be varies widely based on the situation. For most emergency response operations, essential personnel are local and nongovernmental police, fire, and emergency personnel who, in the early stages of an incident, are responsible for protecting and preserving life, property, evidence, and the environment, including emergency response providers as defined in section 2 of the Homeland Security Act of 2002 (6 U.S.C. 101). This includes emergency management, public health, clinical care, public works, and other skilled support personnel (such as equipment operators) who provide immediate support services during prevention, response, and recovery operations. First responders may include personnel from federal, state, local, tribal, or nongovernmental organizations. Table 4.1 describes examples of essential personnel within two categories—first responder personnel and critical infrastructure personnel—followed by a brief discussion of volunteers.

Response Scenarios and Associated Hazards

First responders may be exposed to a variety of hazards upon deployment to a given site. Because hazards vary significantly depending on the scenario, it is important to contemplate both general and specific hazards based on the type of emergency. Figure 4.1 illustrates four response scenarios. Following is a discussion of the possible hazards associated with those scenarios to provide insight regarding probable protective measures necessary. Information is presented based on both acute and chronic hazard potential.

Natural Disasters

Natural disasters present potential hazards that can be both acute and chronic. For responders performing rescue, body retrieval, and infrastructure assessment, acute hazards include impact from moving equipment or tools, flying particles, falling objects, penetration from sharp objects, compression from rollover or rolling

Table 4.1 Essential Personnel

First responder personnel	Personnel who may be classified as first responders include police department/law enforcement, fire department, search and rescue, emergency medical technicians (EMT) and paramedics, emergency managers and staff, emergency medical services/first receivers, National Guard, and logistics (distribution of goods) personnel.
Critical infrastructure personnel	Personnel who may be classified as critical infrastructure personnel include facilities management, solid waste, infrastructure management, engineering, water/sewer, transportation, and government officials/appointees.
Volunteers	Volunteers can provide crucial support, particularly during and after a catastrophic incident. However, it is much more difficult to control the consistency of health and safety knowledge among volunteers, and just-in-time training may be time prohibitive. As a general rule, administrative control of volunteers should be utilized to preclude them from being exposed to hazards. This can be done by assigning volunteers to rear-echelon duties, such as logistics operations. Although logistics operations present hazards that must be mitigated, they are understood in most cases and relatively easy to manage, as opposed to hazards in the actual response zone. However, as needed, specific volunteers serving in an emergency response role may be considered first responders. Specifically, American Red Cross volunteers and others reporting to alternative care facilities and field hospitals, mass dispensing sites, and other locations may fall under the first responder definition, particularly if the hazard is ongoing. Health and safety programming must account for training of volunteers.

objects that can crush feet, and electrical hazards from downed electrical lines and damaged electrical equipment. Heat exhaustion and hypothermia are also potential acute impacts for responders depending on the time of the year and the length of the response.

Acute hazards result from chemical fires involving both stationary and transportation sources including burns and respiratory exposure. Immediately following a building collapse, fire, earthquake, or other catastrophic event, harmful dust and chemical exposure may produce both acute and chronic health effects.

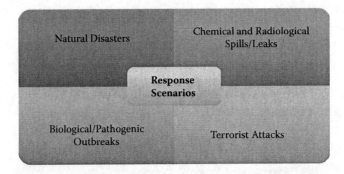

Figure 4.1 The four primary response scenarios.

Chemical and Radiological Exposure

Chemical and radiological exposure can produce both acute and chronic effects. When responding to a chemical or radiological incident, early and specific information regarding source, concentration, and wind direction are crucial for protection of first responders.

Biological/Pathogenic Outbreaks

Biological and pathogenic incidents should be treated as acute until otherwise defined. With respect to biological and pathogenic outbreaks, three primary concerns exist:

1. Is the incident isolated and is PPE useful and warranted?
2. Can first responders be protected with medical countermeasures and are they warranted?
3. Are biosafety (aseptic) procedures warranted to prevent the spread of the pathogen or disease?

   Where the potential for biological exposure exists as a result of natural hazard (i.e., body retrieval), PPE and up-to-date vaccinations may be sufficient to protect responders. When a confirmed outbreak exists or mass prophylaxis will be activated, medical countermeasures for first responders are warranted. For confirmed outbreaks among both animals and humans, biosafety procedures may be warranted to contain the outbreak.

Terrorist Attacks

For terrorist attacks, all of the potential hazards listed earlier may exist and are likely to be presented in a more acute form. Although the acute effects of the events of September 11, 2001, are well understood, studies on first responders from ground zero in New York indicate the onset of a variety of respiratory illnesses due to exposure to dust and chemicals for several days following the attack. The causal factor behind a terrorist attack, whether a conventional bomb, dirty bomb, or chemical weapons, provides a primary indication of acute health effects. However, it is important to consider the potential underlying secondary effects of a large scale attack in protecting first responders.

Protection Programs and Systems

The programs discussed next are largely based on OSHA requirements and hazards that can be anticipated during response to a variety of incidents up to and including large-scale, catastrophic incidents. Although some programs may be more applicable than others, the intent is to provide a comprehensive framework through which a safe response can be facilitated. These programs provide the cornerstone for an effective health and safety program for emergency response operations.

Hazard Communication Program

The purpose of OSHA's Hazard Communication Program is to ensure that, to the extent possible, hazards of all chemicals are known and that information concerning hazards is communicated to potentially affected responders. Information should be provided by means of ongoing and just-in-time responder training programs, which include the following:

1. The contents of the written Hazard Communication Program
2. Container labeling guidelines
3. Material safety data sheets (MSDSs)
4. Discussion of the properties of hazardous chemicals and measures that must be taken to protect responders from these chemicals

The response site must be evaluated for the presence of hazardous chemicals. OSHA regulations require that all chemicals manufactured in or imported to the United States be evaluated for health and physical hazards and MSDSs be supplied to end users. If the response site involves a facility known to have hazardous materials, the hazard evaluation should begin with a review of MSDSs if they are available. Local emergency planning committees (LEPCs) and state emergency response commissions (SERCs) may serve as valuable resources in obtaining MSDSs for a particular site. Building and site identification symbols may also provide information as to the contents of the building or site. If the response site involves residential buildings, office buildings, or a combination of development types, vigilance must

Table 4.2   The Seven Primary Chemical Hazards

Corrosive	Describes a chemical that can cause severe skins burns, for example, battery acid.
Explosive	Describes a chemical that causes a sudden release of pressure, gas, and heat when exposed to sudden shock, pressure, or high temperature.
Flammable	Describes a chemical that catches fire easily with a flash point below 100°F. Some solid materials are also flammable.
Irritant	A chemical that causes swelling and skin rashes on contact.
Radioactive	Describes any material or combination of materials that gives off harmful radiation.
Reactivity	Describes materials that are not stable and can therefore burn, explode, or release dangerous vapors if exposed to heat, air, water, or other particular chemicals.
Toxic	Describes a chemical that can cause illness or death, depending on the chemical and the level of exposure. Many toxic chemicals react with specific body organs. A toxic material enters the body through ingestion, inhalation, absorption through the skin, or injection.

be used in entering buildings or sites with consideration given to household hazardous chemicals. Consideration should also be given to chemicals that are released as a result of burning materials.

Seven primary hazards must be considered when evaluating response sites relative to chemicals. The hazard varies with the level of exposure and the type of chemical. Table 4.2 describes the seven primary chemical hazards.

Routes of entry should also be considered in evaluating chemical hazards. Four routes of entry exist by which a chemical can enter the body.

1. Absorption—Chemicals can pass through the skin or eyes and enter the bloodstream.
2. Inhalation—Breathing chemicals can cause dizziness, nausea, headaches and damage to the lungs or respiratory system. In extreme cases, it can lead to unconsciousness, asphyxiation, or death.
3. Ingestion—Swallowing chemicals is a common means of entry into the body. Responders can be affected if they are doused with chemicals during response activities. Eating or smoking after handling chemicals without washing first increases the risk of chemical ingestion.
4. Injection—If responders puncture their skin accidentally, chemicals on the puncture object can enter the body.

Material Safety Data Sheets (MSDSs)

When available, MSDSs should be used as an integral part of the responder training program on hazardous chemicals, and available MSDSs should be kept in a central location with the safety officer. The National Institute for Occupational Safety and Health (NIOSH) also publishes the *Pocket Guide to Hazardous Chemicals* (*PGHC*), which can be useful when MSDSs are unavailable or as an additional information resource. These resources should be used in training courses to provide responders information regarding chemical hazards.

Given the ad hoc nature of response scenarios, MSDSs may not be available during an incident. However, reporting of chemical use is required by various Environmental Protection Agency (EPA) regulations, and those reports, as well as MSDSs, must be provided to the local LEPC and the SERC. These reports are available to state and local responders and should be used in preincident training to provide responders with a general scope of the hazards present within their jurisdiction. In addition, building, facility, or transportation placards can be used to provide general safety information for initial approach to an incident scene.

MSDSs come in many different forms but contain the same basic information. The main sections are:

Section 1—Chemical Product and Company Identification. This information indicates name of the chemical and the name, address, and phone number of the company that made the chemical.

Section 2—Ingredients. Lists the chemical's hazardous ingredients. Some MSDSs may also include the percentage of each ingredient in the chemical. This section may also give information on how much exposure one can have to the chemical to be within OSHA exposure limits (PEL) or the ACGIH limits (TLV).

Section 3—Hazard Identification. Includes information for emergency situations such as:
- the chemical's normal appearance, odor, and vapors;
- any immediate hazards such as fire, explosion, reactivity, and so forth;
- effects on the body if the chemical enters the body;
- symptoms of overexposure (nausea, headache, etc.); and
- medical conditions that could be made worse by exposure to the chemical.

Section 4—First Aid Measures. Contains information on how to handle first aid and emergency procedures immediately after exposure and before medical help is available. This section also provides information for doctors who may treat the individual.

Section 5—Fire-fighting Measures. Describes the chemical's fire and explosive properties and type of extinguishing material that should be used.

Section 6—Accidental Release Provisions. Describes the necessary actions to take in the event of spill of the chemical or a release to the air. This section also indicates response and clean up procedures.

Section 7—Handling and Storage. Provides information on how to handle the chemical to prevent exposure, hygiene practices to avoid exposure, and storage conditions to avoid with a particular chemical.

Section 8—PPE. Contains information regarding exposure limits; engineering controls, such as ventilation, to protect responders; and PPE such as gloves, goggles, respirators, and so forth.

Section 9—Physical and Chemical Properties. Provides chemical properties of the material such as boiling point, melting point, vapor pressure, solubility, specific gravity, and so forth.

Section 10—Stability and Reactivity. Describes substance stability and how it may react to other substances or conditions, such as heat.

Section 11—Toxicological Information. Contains information regarding health hazards.

Section 12—Ecological Information. Describes how spills or releases may affect ecosystems.

Section 13—Disposal Information. Provides information on proper disposal procedures for a chemical.

Section 14—Transport Information. Contains necessary shipping information required by the Department of Transportation (DOT).

Section 15—Regulatory Information. Refers user to federal, state, and international laws that apply to a chemical.

Section 16—Other Information.

Training is vital, and the training program must ensure that responders understand each section of the MSDS and are able to understand its contents. Understanding the standardized sections will allow responders to refer to necessary information quickly during a response scenario.

Container Labeling

Container labeling is extremely important. Labels may contain information on the potential hazards of the chemicals at the site as well as procedures for their safe use, storage, and handling. Bags, barrels, drums, cylinders, storage tanks, boxes, bottles, and cans must carry labels. Manufacturers and distributors are responsible for labeling each container of hazardous chemicals that they ship. At a minimum, labels should contain the following information about the chemical:

- Identity of the chemical (common and/or chemical name plus any chemical ingredients)
- Name and address of the manufacturer or importer
- Physical hazards
- Health hazards

■ Proper storage and handling procedures (i.e., use only in well-ventilated areas)
■ Protective clothing, equipment, and procedures to work safely with the chemical

All labels are important. Responders should always read the label at any response scene before they move, handle or open a chemical container or initiate any other response measures. Never assume an unlabeled container is harmless. If an unlabeled container or a label that is too torn or worn to read is encountered, precautions must be taken prior to further contact.

If a number of stationary containers are present within a response site, signs may be posted to convey hazard information. Two acceptable labeling methods are the National Fire Protection Association (NFPA) 704, also known as the "fire diamond" or "hazard diamond" (see Figure 4.2) and Hazardous Materials Identification System (HMIS; see Figure 4.3).

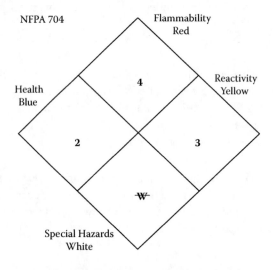

Figure 4.2   NFPA 704 label.

Figure 4.3   HMIS label.

The Importance of Training

Responders must be provided with information and training when they are potentially exposed to hazardous chemicals. Training is required at the time of initial assignment and annually from the date of last training. Training should be held both in the classroom and on the job by a qualified instructor. Training must be documented and records must be maintained. Retraining is required when new hazards are introduced to the jurisdiction. Just-in-time training will be performed to augment annual training when response conditions warrant explanation of hazards pursuant to the event.

At this time, unlike chemical hazards, standardized MSDS requirements for pathogens do not exist in the United States. However, it is important to provide the similar information to those who may be exposed to pathogens in the course of response actions. Two sources of information exist that provide standardized information regarding pathogens suitable for use in training and response. The Public Health Agency of Canada produces MSDSs for pathogens, which are available at www.phac-aspc.gc.ca/msds-ftss/. The North American Treaty Organization (NATO) has also produced Clinical Data Sheets for Selected Biological Agents available at www.fas.org/nuke/guide/usa/doctrine/dod/fm8-9/2appb.htm.

Personal Protective Equipment (PPE) Program

OSHA regulations require PPE for eyes, face, head, and extremities for hazardous tasks. Other considerations include hearing protection, protective clothing, protective shields and barriers, and respiratory equipment. Hazards may include those of processes or environment, chemical hazards, radiological hazards or mechanical irritants that may cause injury or impairment through absorption, inhalation, or physical contact. In the spring of 1994, OSHA revised its PPE regulation to require a hazard assessment to determine where PPE should be used to prevent potential personal injury (Figure 4.4).

In addition to OSHA standards and other consensus standards, the U.S. Department of Homeland Security (DHS) has adopted several standards developed by the National Fire Protection Association (NFPA) relating to PPE for first responders:

NFPA 1851, Standard on Selection, Care and Maintenance of Structural Fire Fighting Protective Ensembles (2001 Edition)—This document specifies minimum selection, care, and maintenance requirements for structural fire fighting protective ensembles, and the individual ensemble elements that include coats, trousers, coveralls, helmets, gloves, footwear, and interface components that are compliant with NFPA 1971. This first edition of NFPA 1851, issued in 2001, was developed to be a companion document for NFPA 1971, Standard on Protective Ensemble for Structural Fire Fighting.
NFPA 1852, Standard on Selection, Care, and Maintenance of Open-Circuit Self-Contained Breathing Apparatus (SCBA) (2002 Edition)—This standard

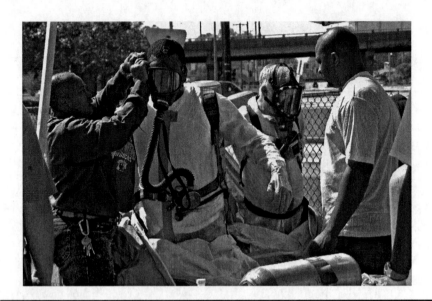

Figure 4.4   Donning PPE. (Courtesy FEMA, Win Henderson.)

establishes procedures as part of a program to provide care and maintenance for open-circuit SCBA and combination SCBA/SAR in order to reduce the safety risks and potential health risks associated with poorly maintained, contaminated, or damaged SCBA. The edition of NFPA 1852 issued in 2002 was the first.

NFPA 1951, Standard on Protective Ensemble for Technical Rescue Operations (2007 Edition)—Based on work begun in 1997, this standard answers the need for PPE for fire and emergency services personnel operating at technical rescue incidents involving building or structural collapse, vehicle accidents, confined spaces, trench cave-ins, scaffolding collapses, high angle climbing accidents, and similar incidents. The first edition of NFPA 1951 was issued in July 2001.

NFPA 1971, Standard on Protective Ensembles for Structural Fire Fighting and Proximity Fire Fighting (2007 Edition)—Based on work begun in 1973, this standard establishes minimum levels of protection for firefighting personnel assigned to fire department operations including but not limited to structural fire fighting, proximity fire fighting, rescue, emergency medical, and other emergency first responder functions. The first edition of NFPA 1971 was issued in 1975.

NFPA 1975, Standard on Station/Work Uniforms for Fire and Emergency Services (2004 Edition)—Based on work begun in 1975, this standard specifies requirements for the design, performance, testing, and certification of nonprimary protective station/work uniforms and the individual garments comprising station/work uniforms. The document sets requirements for fire and emergency services personnel station/work uniforms that will not contribute to burn injury severity. The first edition of NFPA 1975 was issued in 1985.

NFPA 1981, Standard on Open-Circuit Self-Contained Breathing Apparatus (SCBA) for Emergency Services (2007 Edition)—Based on work begun in 1975, this standard specifies the minimum requirements for the design, performance, testing, and certification of open-circuit self-contained breathing apparatus (SCBA) and combination open-circuit self-contained breathing apparatus and supplied air respirators (SCBA/SAR) for the respiratory protection of fire and emergency responders where unknown, immediately dangerous to life and health (IDLH), or potentially IDLH atmospheres exist. The first edition of NFPA 1981 was issued in 1981.

NFPA 1982, Standard on Personal Alert Safety Systems (PASS) (2007 Edition)—Based on work begun in 1980, this standard was developed in response to requests from the fire service to establish requirements for a device that would sound an audible signal for aid if a firefighter became incapacitated while operating at an emergency. This standard specifies minimum requirements for the design, performance, testing, and certification for all Personal Alert Safety Systems (PASS) for emergency services personnel. The first edition of NFPA 1982 was issued in 1983.

NFPA 1991, Standard on Vapor-Protective Ensembles for Hazardous Materials Emergencies (2005 Edition)—Based on work begun in 1986, this standard specifies the minimum requirements for the design, performance, testing, and certification of vapor-protective ensembles and individual protective elements for chemical vapor protection for fire and emergency service personnel. Additional optional criteria are provided for ensembles and individual protective elements that provide protection for chemical flash fire escape, liquefied gas, chemical and biological warfare agents, and chemical and biological terrorism incidents. The first edition of NFPA 1991 was issued in 1990.

NFPA 1992, Standard on Liquid Splash-Protective Ensembles and Clothing for Hazardous Materials Emergencies (2005 Edition)—Based on work begun in 1985, this standard specifies minimum requirements for the design, performance, testing, documentation, and certification for liquid splash-protective ensembles, ensemble elements, and protective clothing used by emergency response personnel during hazardous materials incidents. The first edition of NFPA 1992 was issued in 1990.

NFPA 1994, Standard on Protective Ensembles for First Responders to CBRN Terrorism Incidents (2007 Edition)—Based on work begun in 1998, this standard specifies the minimum requirements for the design, performance, testing, and certification of protective ensembles for fire and emergency services personnel operating at domestic terrorism incidents involving dual-use industrial chemicals, chemical terrorism agents, or biological terrorism agents. The intent is that the ensembles would be available in quantity, easily donned and used, and designed for single exposure use. The first edition of NFPA 1994 was issued in 2001.

NFPA 1999, Standard on Protective Clothing for Emergency Medical Operations (2003 Edition)—Based on work begun in 1990, this standard specifies the minimum documentation, design, performance, testing, and certification requirements for new single-use and new multiple-use emergency medical protective clothing, including garments, gloves, footwear, and face protection devices, used by fire and emergency services personnel during emergency medical operations. The purpose of the standard is to establish a minimum level of protection from contact with blood and body fluid-borne pathogens for personnel performing patient care during emergency medical operations. The first edition of NFPA 1999 was issued in 1992.

NFPA 2112, Standard on Flame-Resistant Garments for Protection of Industrial Personnel Against Flash Fire (2007 Edition)—Based on work begun in 1999, this standard shall provide minimum requirements for the design, construction, evaluation, and certification of flame-resistant garments for use by industrial personnel, with the intent of providing a degree of protection to the wearer and reducing the severity of burn injuries resulting from accidental exposure to hydrocarbon flash fires. The first edition of NFPA 2112 was issued in 2001.

NFPA 2113, Standard on Selection, Care, Use, and Maintenance of Flame-Resistant Garments for Protection of Industrial Personnel Against Flash Fire (2007 Edition)—Based on work begun in 1999, the purpose of this standard is to reduce the health and safety risks associated with the incorrect selection and use of flame-resistant garments and those risks associated with incorrectly maintained, contaminated, or damaged flame-resistant garments. The first edition of NFPA 2113 was issued in 2001.

These standards provide consistency among various forms of PPE and provide useful guidelines in developing specifications for purchasing PPE for first responders.

Responsibilities of the Safety Officer

To effectively protect responders, the safety officer should perform the following tasks upon arrival at a response scene:

1. Assess work to be performed by responders to determine if hazards are present that require use of PPE.
2. Select appropriate types and sizes of PPE to protect responders from potential dangers identified during assessment and require each responder to use appropriate PPE for the tasks at hand.
3. Communicate selection decisions to each affected responder.

4. Certify, in writing, that the work site hazard assessment has been completed that identifies areas evaluated, findings, person certifying completion of the assessment, and the date of assessment.
5. Provide training to each responder required to use PPE and ensure that each responder can both properly use PPE and demonstrate an understanding of when PPE is necessary; what PPE is necessary; how to properly don, doff, adjust, wear, and otherwise use PPE; limitations of PPE; and proper care, maintenance, useful life, and disposal of PPE.
6. Provide retraining when changes in the environment render previous training obsolete, changes in types of PPE to be used render previous training obsolete, or inadequacies in an affected responder's knowledge or use of assigned PPE indicate that the responder has not retained understanding or skills necessary to effectively use PPE.
7. Keep a training log that certifies that each affected responder has received and understands required training including name of each responder trained, date(s) of training, and subject of certification

A walk-through assessment of the incident site, facility, or tasks is the best method to determine the sources of hazards to responders. The following basic hazard categories should be considered:

■ Impact (from moving equipment or tools, flying particles, falling objects, etc.)
■ Penetration (from sharp objects)
■ Compression (rollover, rolling objects that can crush feet)
■ Chemicals (exposures to)
■ Heat (burns, ignition of protective equipment, eye injury)
■ Harmful dust
■ Light (optical)
■ Radiation (welding, brazing, etc.)
■ Electrical

After the assessment, all information should be analyzed to ensure proper selection of PPE for the hazards involved. Hazards at the event site should be compared with the PPE available, and PPE should be selected that provides a level of protection above what is required to protect responders from hazards. It is possible that a responder will be exposed to several hazard sources at the same time, so it is important to consider all hazards when selecting PPE for a specific area or task. PPE must be provided in sizes necessary to fit all exposed responders. Improperly fitted PPE may negate protective properties and make it uncomfortable for the user. Responders must be informed regarding proper care and use of PPE and must be made aware of manufacturer's warning labels and limitations of PPE they are using.

Respiratory Protection Program

Respirators can play a critical role in protecting responders from toxic hazards. Standard operating procedures should be established for safe use of respiratory protective devices. Respirators should be used during emergencies where the exposure risk is either unknown or cannot be reduced by means other than the use of respiratory protection.

OSHA standard 29 CFR 1910.134 requires establishment of a respiratory protection program manager, who will make determinations of whether respirators are needed for a particular site or task. Air monitoring or visual inspection of the task in question is necessary to make this determination. This determination should be based on the likelihood of responder exposure to a particular air contaminant at or above the OSHA permissible exposure limit, the threshold limit value established by the American Conference of Governmental Industrial Hygienists (ACGIH), or other recognized standard. NIOSH, OSHA, and ANSI (American National Standards Institute) safety guidelines should be used to evaluate the hazards and select the appropriate equipment. Only respiratory protection equipment approved by the NIOSH/Centers for Disease Control and Prevention should be used. Responders who are required to use respirators should be properly fitted, fit-tested, medically screened, and trained.

Responders may choose to wear respirators even in locations where it is deemed that respirator use is not required due to exposure to a particular hazard. Under these circumstances, the program manager may provide respirators at the request of responders or permit responders to use their own respirators if he or she determines that such respirator use will not in itself create a hazard. If voluntary respirator use is permitted, the program manager shall provide the respirator users with voluntary use respirator information, distributed by OSHA. In addition, the responder is still subject to medical evaluation to ensure that he or she is medically able to use that respirator. The program manager must also ensure that the respirator has been cleaned, stored, and maintained so that its use does not present a health hazard to the user. These requirements do not apply to responders whose only use of respirators involves the voluntary use of filtering face pieces (dust masks).

The program manager is responsible for respirator selection. Selection criteria will be determined for each response, based on the nature and type of hazard, prior to entry by response personnel into the hazard zone. The program manager, upon request, will provide guidance on respirator selection for specific hazards. Respiratory hazards and relevant site and user factors must be included in selecting respirators.

Two general types of respirators are available for use: (1) air-purifying respirators, which use cartridges or particulate air filters to remove contaminants from the breathing zone; and (2) air-supplying respirators, which use supplied air from compressed air tanks, SCBA, or airline systems. In selecting the proper respirator for given circumstances, several factors should be considered:

1. Nature of the hazard—The nature of the hazard will be identified to ensure that overexposure does not occur. Relevant factors include oxygen deficiency, physical and chemical properties of the hazard, physiological effects on the body, concentrations of the toxic substances, the PELs, and warning properties. This also includes information regarding the type of chemical properties and filtering capabilities of the respirator filter media.
2. Nature of the hazardous operation—The program manager will be familiar with the operations that require responders to use respiratory devices. These include equipment, processes and process characteristics, work area characteristics, materials used or produced during the process, the responder's duties and actions, and any unusual characteristics that may necessitate alternate respirator selection.
3. Location of the hazardous area—The location of the hazardous area must be identified and a backup system planned, if necessary. Respirable air locations must be known before entry is made into a hazardous area so escape or emergency operations may be planned. All entry into confined spaces either requiring respiratory protection or backup respiratory protection will first be identified and the appropriate notifications and permits completed.
4. Time of respiratory protection—The length of time a respirator will be worn by a responder shall be evaluated. This is most pronounced when using an SCBA in which the air supply is limited. Time is also a factor during routine use of air-purifying respirators when the responder's breathing and comfort become affected by a clogged filter cartridge or when the duration will cause undue stress on the responder. Cartridges may also become loaded and ineffective after a period of time. Refer to the respirator's end-of-service-life indicator (ESLI) to determine when the respirator is no longer functioning properly.
5. Responder health—Effective use of a respirator depends on an individual's ability to wear a respirator, as determined by a physician. Most respiratory devices increase physical stress on the body, especially the heart and lungs. Before beginning an assignment that requires use of a respirator, responders must undergo a medical evaluation to determine that they are capable of wearing a respirator for the duration of the assignment.
6. Nature of assignment—Work activities that will be performed while wearing a respirator have an integral impact on respirator selection. Selection of the proper respirator shall be evaluated to determine the type that causes the least amount of disruption to the task yet provides the desired protection.
7. Protection factors (PF)—The protection afforded by respirators depends on the seal of the face piece to the face, around valves and through or around cartridges or canisters. Depending on these criteria, the degree of protection may be ascertained and a relative safety factor assigned. Protection factors are only applicable if all elements of an effective respirator program are in place and being enforced. The PF is multiplied by the exposure limit of the

chemical agent or particulate to determine the maximum concentration at which the respirator will be effective. The protection factors are subject to change.

- Half-face piece respirators: 10 PF
- Full-face piece respirators: 50 PF
- Self-contained breathing apparatus: 10,000 PF

8. Limitations—Air purifying respirators cannot be used in oxygen-deficient atmospheres, in IDLH atmospheres, or in confined spaces with unknown or known hazardous atmospheres. They can only be used for protection against contaminants listed on each cartridge or identified by the manufacturer and approved by NIOSH. The wearer should leave an area immediately if odors are detected inside the mask or if the breathing resistance increases. Half-mask respirators shall not be worn with contact lenses or when facial hair or anything else extends under the sealing surface.

Chemical cartridges for air purifying respirators will not provide protection for certain chemicals, illustrated in Figure 4.5. Air supplying respirators are reserved for emergency use only. Their use is restricted to those responders trained, fitted, and medically cleared to wear an SCBA.

Respirators should be issued to responders only if the responder has received appropriate medical evaluation, fit testing, and training. Responder training should occur at the time of initial employment or when the responder is assigned to a particular task requiring use of a respirator. The extent and frequency of responder training depends primarily on the nature and extent of the hazard. At a minimum, all responders assigned to a task that requires a respirator will be trained in basic respirator practices initially upon assignment and annually thereafter.

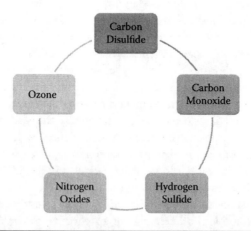

Figure 4.5   Chemicals not safe for APRs.

Hearing Conservation Program

The purpose of the Hearing Conservation Program is to help protect responders' hearing, prevent permanent hearing impairment, avoid negative psychological effects, and avoid speech interference. Although situations requiring a hearing conservation program in response scenarios may be rare, it is important to reduce hazardous noise where feasible, to provide and require use of hearing protection when applicable, to provide responder education, and to maintain a complete and comprehensive hearing conservation program. This program is written in accordance with OSHA regulation 29 CFR 1910.95.

The safety officer is responsible for maintaining this program at the response scene. Responsibilities include implementing this program, conducting noise surveys, and providing hearing protection. Responders are responsible to actively participate in the hearing conservation program and to use hearing protection devices properly where required.

Implementing this program to achieve maximum benefit requires that much of the framework is instituted before an incident occurs. It is also important that as many agencies and jurisdictions as possible follow the guidelines set forth herein so that all responders understand this program and are able to respond appropriately when an incident occurs. In many cases during a response scenario, it may be impractical to perform some of the tasks set forth herein, such as noise monitoring. As a rule of thumb, the threshold for wearing hearing protection is probably exceeded if activities produce enough noise on a continuous basis that people have to raise their voices to be heard. In addition, responders from multiple jurisdictions may or may not have baseline audiograms (hearing test results in graph or picture form). In this case, just-in-time training regarding hearing protection should be performed and audiograms should be scheduled as soon after the response as possible.

Responders exposed to noise at or above an eight-hour time weighted average (TWA) of 85 decibels on the A scale (dBA) will have annual training. Information provided in training will be updated to be consistent with changes in protective equipment and work processes. Responder training shall include the following information:

- Effects of noise on hearing
- Purpose of hearing protectors, advantages, disadvantages, and attenuation of various types, and instructions on selection, fitting, use and care
- Purpose of audiometric testing and an explanation of test procedures.

Objectives of noise monitoring at the response site include identifying individual responders who are to be participants in the hearing conservation program due to their exposure to noise at or above 85 decibels, and obtaining sufficient noise exposure information to provide appropriate selection of adequate hearing protection.

When information indicates exposure may equal or exceed an eight-hour time-weighted average of 85 dBA, monitoring should be conducted. Monitoring can be conducted using either a sound level meter or a dosimeter (a device that screens noise levels). If circumstances dictate the need for personal sampling, such as high worker mobility, significant variations in sound level, or a significant component of impulse noise, a dosimeter should be used. All continuous, intermittent, and impulsive sound levels from 80 dBA to 130 dBA will be integrated into noise measurements. Instruments used in this sampling will be calibrated to ensure measurement accuracy.

As stated previously, it may be impractical to perform some of the tasks set forth herein such as noise monitoring. As a rule of thumb, the threshold for wearing hearing protection is probably exceeded if activities produce enough noise on a continuous basis that people have to raise their voices to be heard.

Hearing protection devices are designed to prevent intrusion of unwanted sound and possible damaging sound into the ear canal. Two major forms of hearing protection are earplugs and earmuffs. Hearing protection devices will be provided for affected responders exposed to an eight-hour time-weighted average of 85 dBA.

The responder's audiogram will be compared to the responder's baseline to determine if the audiogram is valid and if a standard threshold shift (STS) has occurred. An STS is a significant change in hearing ability. Additional testing may be conducted within thirty days after the annual audiogram to determine if the first annual audiogram is valid, and the second testing results may be used. If an STS is identified, the following steps must be taken:

1. Responders currently not using hearing protection must be fitted with an adequate hearing protection device, trained in its use and care, and must be required to wear the hearing protection in high noise areas.
2. Responders experiencing an STS who are already using hearing protection must be refitted and retrained in the use of hearing protection and offered hearing protection that offers greater attenuation, if necessary.
3. Responders must be referred for a clinical audiological evaluation or an otological examination if additional testing is necessary or if the jurisdiction suspects that a medical pathology of the ear is caused or aggravated by wearing hearing protection devices.
4. Responders must be informed of the need of an otological examination if a medical pathology of the ear that is unrelated to the use of hearing protection devices is suspected or found.
5. STSs of 25 dB from the responder's baseline should be recorded and placed in the responder's medical file. A medical evaluation should also be sought to identify appropriate measures to restore hearing or prevent further loss.

An annual audiogram may be substituted for the baseline audiogram when, in the judgment of the audiologist, otolaryngologist, or physician who is evaluating the audiogram, the STS revealed by the audiogram is persistent or the hearing threshold shown in the annual audiogram indicates significant improvement over the baseline audiogram. Substitution of the baseline audiogram will ensure that the baseline reflects the actual hearing threshold to the extent possible and that the same shift is not repeatedly identified.

Records obtained from noise exposure measurements must be retained for at least two years. Audiometric test records must be retained for the duration of the affected responder's employment and an additional thirty years. Records may be accessed upon request of the responder, former responders, and representatives designated by the individual responder.

Bloodborne Pathogens Program

The purpose of the Bloodborne Pathogens Program is to provide methods to safeguard responders at an incident scene from exposure to blood and other potentially infectious materials. As determined through exposure assessment, this program applies to individuals who, in an emergency situation, have the potential for being exposed to blood or other potentially infectious materials when responding to injuries or resulting from response scene incidents. This program should also apply to personnel directly responsible for cleanup of an incident site after an accident has occurred. Exposure determinations should be made without regard to use of personal protective devices (responders are considered to be exposed even if using PPE).

The safety officer is responsible for implementing all elements of this program for responders and may designate staff members to assist in implementation, maintenance, and administration of the program. Responders are responsible for the following:

- Knowing what tasks they perform, if any, that may lead to occupational exposure
- Attending bloodborne pathogens training and refresher courses
- Planning and conducting all operations in accordance with work practice controls in this program
- Developing and using good personal hygiene habits

Responders shall be advised of this program and its location during orientation and other education and training sessions. The program should be reviewed and revised annually or when new procedures or tasks are implemented that could affect responder exposure, when procedures or tasks are revised that could affect responder exposure, or when new functional positions are established that may involve exposure.

Biohazard labels (see Figure 4.6 for examples) are the most obvious warnings of possible exposure to bloodborne pathogens. Approved labels or red color-coded containers will be used when necessary. The safety officer is responsible for ensuring that appropriate labels are used. These symbols or similar labels shall be used on all disposal bags, contaminated equipment, and infection control kits.

Figure 4.6 Biohazard labels.

All responders who have potential for exposure to bloodborne pathogens must attend a comprehensive training program designed to give responders the information needed to protect them from exposure. New responders or responders changing jobs or job functions requiring training in bloodborne pathogens will receive this training at the time of their new job assignment. After initial training, responders will attend annual refresher courses. Training may occur sooner if substantive changes in job descriptions take place.

During training, responders will receive:

- Information regarding the OSHA Bloodborne Pathogens Standard, 29 CFR 1910.1030
- A copy of the exposure control plan
- Information regarding bloodborne diseases (epidemiology, symptomology, modes of transmission)
- Information about identifying tasks that may involve exposure to blood and other potentially infectious materials
- Information regarding use and limitations of exposure control mechanisms, including engineering controls, work practice controls, PPE, and personal hygiene
- Information regarding types, selection, use, location, removal, and handling of PPE
- Information regarding hepatitis B vaccines, including efficiency, safety, mode of administration, and benefits
- Instructions regarding actions taken upon exposure (reporting, medical follow-up, counseling)
- Visual warnings of biohazards including labels, signs, and color-coded containers

Responders will also be provided an opportunity to ask the instructor questions. The bloodborne pathogen training program will consist of a classroom atmosphere and/or computer-based training.

The safety officer will maintain training records. The records will contain dates and contents of training sessions, names of the instructors, and names and job titles of the responders attending the training sessions. Training records shall be available for examination and photocopying by responders, their representatives, and state authorities.

Universal Precautions for Bloodborne Pathogens

Universal precautions should be observed at all times to prevent contact with blood or other potentially infectious materials. All blood or other potentially infectious material should be considered infectious regardless of perceived status of source individual. As such, these steps must be taken:

Gloves should be worn when touching blood or other body fluids, mucus membranes, or nonintact skin or when handling items or surfaces soiled or reasonably suspected to be soiled with blood or other body fluids. Gloves should be disposed of after a single use.

If it is anticipated that droplets of blood or any body fluid may come in contact with the mucus membranes of the responder's eyes, nose, or mouth, he or she should wear protective equipment such as goggles or a face shield.

Hands and other skin surfaces should be washed immediately if contaminated with blood or other body fluids. Hands should also be washed upon removal of gloves.

Any contaminated items such as razors, knife blades, broken glass, or equipment should be disposed of in a puncture and leak-proof container labeled for disposal of such items.

To minimize exposure to bodily fluids during CPR, nonreflexive breathers or other disposable aids should be used. Reusable breathers can be used only when approved decontamination procedures have been used.

If clothing is contaminated, it must be removed as soon as possible.

Eating, drinking, applying cosmetics or lip balm, and handling contact lenses are prohibited in rooms used for first aid purposes.

Engineering controls can be used to eliminate or minimize responder exposure to bloodborne pathogens. Responders must use containers and appropriate disposal bags for potentially infectious waste. Responders who are exposed to blood and other potentially infectious materials must use hand-washing facilities.

Infection control kits must be kept in areas where responders may be exposed to bloodborne pathogens. The infection control kit shall contain the following items:

■ Three pairs of puncture-resistant gloves
■ Antibacterial wipes
■ Antibacterial cleanser
■ Dry, nonporous wipes
■ Two red biohazard bags

PPE is the last line of defense against bloodborne pathogens. PPE necessary to protect responders against exposure should be provided. This equipment should be provided and may include gloves, safety glasses, goggles, face shields, and respirators. All responders and work areas must have appropriate PPE available and accessible. PPE should be chosen based on anticipated exposure to blood or other potentially infectious materials. To ensure that PPE is not contaminated and in appropriate working condition, safety officers and responders must ensure that:

- All PPE is inspected periodically and repaired or replaced as needed;
- All reusable equipment is cleaned, laundered, and decontaminated as needed at no cost to responders;
- All potentially contaminated PPE is removed prior to leaving a work area;
- Disposable gloves are replaced as soon as practical after contamination or if they are torn, punctured, or otherwise lose their ability to function as a barrier to exposure;
- Reusable utility gloves are assigned to each responder as needed and are decontaminated after each use;
- Protective clothing is worn whenever potential exposure to the body is anticipated.

Uniforms are not considered protective clothing. Any time these garments are penetrated by blood or any other potentially infectious materials, the item or items should be removed immediately or as soon as feasible. If contaminated, these items should be disposed of and replaced.

Procedures for Decontamination

Decontaminating oneself is only part of decontamination. Maintaining facilities in a clean and sanitary condition is an important part of the Exposure Control Plan for bloodborne pathogens. The following are practices of decontamination:

1. Wearing disposable gloves, the responder will use paper towels to remove as much visible material as possible.
2. The item will then be cleaned using a solution of ten parts water to one part bleach (10:1). Antibacterial cleansers rated for use with blood or body fluids may be substituted at the safety officer's discretion.
3. Any surface or equipment contaminated with blood or body fluids will be cleaned as soon as possible.
4. Responders wearing disposable or utility gloves will use paper towels to remove visible material and then decontaminate using 10:1 bleach solution. An antibacterial cleanser rated for use with blood or body fluids may be substituted at the discretion of the safety officer.

5. Cleaning products, such as paper towels and gloves, must be placed in plastic bags and burned. If an incinerator is not available, bags will be removed by contract with an appropriately qualified vendor. Disposal bags will be red in color and marked with the biohazard label.
6. Regulated wastes including bandages, feminine hygiene products, and so forth, are to be placed in biohazard bags for disposal consistent with the aforementioned manner.

Hepatitis B Vaccination Program

To protect responders from possible hepatitis B infection, vaccinations should be offered to those determined to have potential for exposure to bloodborne pathogens. The vaccination will be made available within ten working days of job assignment or ten days after exposure. Responders who decline the vaccine must sign a form. Responders who initially decline but later wish to have the vaccine may request and receive it within ten days.

The safety officer will schedule vaccinations and maintain documentation accordingly. Copies of responder consent or refusal forms must be kept in files on site. A copy of consent or refusal forms and a copy of the vaccination record will be placed in the confidential medical section of the responder's master file. Vaccinations shall be performed under supervision of licensed physicians/healthcare professionals.

Postexposure Evaluation and Follow-Up

If a responder is involved in an incident where exposure to bloodborne pathogens may have occurred, two actions will be taken immediately:

1. The safety officer will ensure that the responder receives medical consultation and treatment as quickly as possible.
2. An investigation of circumstances surrounding the exposure incident will ensue in order to evaluate the situation and determine potential changes in procedure. Investigations will be initiated within twenty-four hours of the incident; an exposure incident investigation form must be filled out to gather the following information:
   - Details such as where, when, and how the incident occurred
   - Type of potentially infectious materials involved
   - Source of the infectious materials
   - Circumstances surrounding the incident
   - PPE being used at the time
   - Action taken as a result of the incident

The follow-up process consists of the following steps:

1. The responder is provided with documentation regarding the routes of exposure, the circumstances under which the exposure occurred, and identification of the source individual (if possible).
2. The blood of the source individual is tested (if possible) to determine HBV and/or HIV infection. This information will also be made available to the exposed responder if it is obtained. At that time, the exposed responder will be made aware of any applicable laws and regulations concerning disclosure of the identity and infectious status of the source individual.
3. The blood of the exposed responder is collected and tested for HBV and HIV status.

Once these procedures have been completed, an appointment is arranged with a qualified healthcare professional to discuss the medical status of the exposed responder. This includes an evaluation of any reported illnesses as well as any recommended treatment. Much of the information involved in this process must remain confidential. The safety officer is obliged to do everything possible to protect the privacy of all of the individuals involved.

Information Provided to the Healthcare Professional

To assist in the evaluation, the safety officer will forward a number of documents to the healthcare professional, including:

- A copy of the OSHA Bloodborne Pathogen Standard
- A copy of the exposure incident investigation form and any information describing exposure incident
- The exposed responders' relevant medical records
- Any other pertinent information

Healthcare Professional's Written Opinion

Within fifteen days of the consultation, the healthcare professional will provide to the safety officer a written opinion evaluating the exposed responder's medical status. The safety officer will then notify the exposed responder of the results of that evaluation. The healthcare professional shall limit written correspondence to issues relevant to the exposure incident and shall inform the responder of the results of the evaluation and of medical conditions resulting from exposure to blood or other potentially infectious materials.

Medical Records

To ensure sufficient medical information is available, the safety officer shall maintain responder medical records in a confidential manner. Medical records will not be disclosed to anyone without responder's written consent, except as required by law. Information to be kept includes:

- The name of the responder
- The responder's social security number
- A copy of responder's hepatitis B vaccination status
- Copies of results of examinations, medical testing, and followup procedures that took place as a result of a responder exposure to bloodborne pathogens
- Copies of information provided to consulting healthcare professional due to bloodborne pathogen exposure.

Confined Space Entry

Responders may be deployed to a scene involving confined spaces. Confined spaces may present toxic, flammable, or oxygen-deficient hazards or a host of other hazards that require special consideration and care while performing response duties. These requirements apply to all responders working in and around identified confined spaces during a response scenario. The program should be designed to maintain the safety of the responder entering the space through a written permit and atmospheric monitoring in compliance with OSHA 29 CFR 1910.146. A brief description of confined space entry program procedures is provided in this section.

The purpose of the Confined Space Entry Program is to ensure safe entry into tanks, pits, and other confined spaces. By following this program, the hazards associated with confined space entry can be controlled or eliminated. The procedures will enable trained responders to identify a confined space including the characteristics that make the confined space hazardous, perform the necessary actions to eliminate or reduce the hazards to a safe level for entry, test the environment within the confined space with monitoring equipment to ensure that it is safe before entering, use the necessary safety equipment and protective clothing needed for safe entry, conduct a safe entry after all hazards have been eliminated or reduced to a level safe for entry, minimize the development of hazards inside the confined space while work is being done within the space, properly use the permit system required for all confined space entries; perform attendant duties, and conduct emergency rescues.

Before entry, the entry supervisor must check the permit and sign it certifying that the conditions are safe for entry. Ladders, boatswain chairs with lifeline, or other safe means are used for safe access into and out of confined spaces. The entrant(s) must check to ensure that they have all of the necessary safety equipment needed for entry. The entrant(s) must have a full body harness, which is attached to a lifeline; the lifeline must be attached to a winch or other retrieval system.

The authorized entrants may enter the confined space without special protective safety or breathing equipment when the following conditions exist in the space:

- The oxygen content is between 19.5 percent and 23.5 percent. If the oxygen content is too high, it can be purged with fresh air to lower it to a safe level. High oxygen levels and a source of ignition will lead to an explosion—and probable death—inside the confined space.
- The level of a flammable substance is below 10 percent of its lower explosive limit.
- No contaminant exists in the confined space at a level above its permissible exposure limit. If it does, full-face respirators with the correct cartridges for the contaminant must be worn. If a level of a contaminant exists that exceeds the capacity of the full-face respirator and cartridges, an air supplied respirator with a backup supply or an SCBA must be worn by the entrant(s).

Absolutely no entry is allowed if oxygen levels remain above 23.5 percent. Purging with fresh air will readily drop the level. The space must not be entered if the concentration of airborne combustible dust obscures vision at a distance of five feet or less. Continue to ventilate with fresh clean air until safe for entry. Authorized entrants must wear goggles and respirators with HEPA cartridges in this case.

Control of Hazardous Energy (Lockout/Tagout)

This program is established to prevent injury to responders or damage to equipment from unexpected startup or a release of energy during a response scenario. It may apply to electrical, hydraulic, or other forms of energy that can cause injury or death to responders. This program establishes procedures for responders to use before performing such tasks as initial damage assessment, search and rescue, and recovery. In this program, equipment refers to electrical, hydraulic, or mechanical installations present at the scene that store energy, which can cause injury due to inadvertent start up. This program is useful in controlling hazards following catastrophic events where electrical hazards may be present.

The safety officer will be responsible for implementing this program at a response scene. All affected responders will be trained and retrained whenever there is a change in work operations or equipment. The safety officer will perform periodic inspections to ensure that procedures are followed and to prevent injury. Responders must comply with all lockout/tagout procedures. Whenever questions arise regarding lockout/tagout, responders should consult the safety officer.

Only authorized responders may perform lockout procedures. An authorized responder has complete knowledge of the lockout/tagout program and is the person who actually performs the lockout of the equipment. A list of authorized responders will be maintained by the safety officer.

An affected responder is a responder whose job requires him or her to work around machinery or equipment on which servicing or maintenance is being performed under this procedure, or whose job requires him or her to work in an area where such servicing or maintenance is being performed. These responders must be able to recognize lockout/tagout devices immediately, know the purpose of the devices, know not to disturb the devices, and be aware of the danger of violating or disregarding procedures. The following general lockout/tagout procedures will apply to all response operations where there is a threat of uncontrolled release of energy.

All equipment/locations must be locked and tagged by each responder involved in the response operation to protect against accidental or inadvertent operation when such operation could cause injury to personnel or damage to equipment. Sources of energy, such as springs, air, and hydraulics must be evaluated in advance to determine whether to retain or relieve the pressure prior to starting the work.

Locks are for the personal protection of responders and are only to be used for locking out equipment. Personal locks will contain a brass (or other suitable material) tag clearly marked with the responder's name.

Each lock will have only one key. This key will be kept with the lock when it is stored. When using a lock for the lockout procedure, each responder must keep the key in his or her possession.

Responders must request assistance from the safety officer if they do not know where or how to lockout/tagout equipment. Any questions concerning the lockout/tagout procedure should be directed to the safety officer.

Tagout without lockout is to be allowed only when there is no means to lock out the equipment. All procedures for the lockout must be followed. However, instead of affixing the lock, a tag must be used that indicates:

- The identity of the person who applied it
- The state of the hazard if operated
- The statement, "Do not start, do not open, do not energize, or do not operate"

In addition to tagging the equipment, other safety measures must be taken when appropriate. These measures could include removal of a fuse, blocking a controlling device, or removing a valve handle. Each responder working on the equipment will have personal tags. No one can remove a tag except the person who applies it. If there is a change in shift, the incoming responders shall affix their own tag devices and the outgoing responders should remove their tags.

If more than one authorized responder is working on a machine or equipment, each involved responder must apply his or her lockout device. Additionally, the responsibility for the multiple lockout situation must be assigned to a single, authorized person for a set number of responders working under the protection of the group lockout/tagout device. He or she must coordinate the affected work force, ensure that continuity of protection will be maintained during the lockout, and be the last person to remove his or her lock when the work is complete.

A machine or tool connected to a plug-in cord is considered effectively controlled when the machine's connecting plug is unplugged from its receptacle and the responder performing work on the equipment has exclusive control of the plug. The plug is under exclusive control if it is physically in the possession of the responder or in arm's reach and in the line of sight of the responder. If this cannot be accomplished, a locking device must be applied to the plug on the machine's power cord. If stored energy is present in the system, it must be relieved or restrained before starting work on the equipment.

Lockout devices may only be removed by the responder who originally placed the lock on the machine. However, certain instances may dictate that the responder's lock may need to be removed by another responder. This is allowable, provided that the device is removed under the direction of the safety officer and that the following steps are taken to protect the original responder:

- Verify the authorized responder who applied the lockout device is not at the scene;
- Make all reasonable efforts to contact the authorized responder to inform him or her that his or her lockout/tagout device has been removed;
- Ensure the authorized responder has this knowledge before he or she resumes work at the scene.

These efforts must be documented and kept on file for one year for the purposes of program review.

Each *authorized* responder will receive training to provide complete knowledge of energy control procedures that includes, but is not limited to:

- Hazardous energy recognition
- Type and magnitude of energy present
- Types and quantities of energy control devices
- Isolation
- Points of control
- Lockout/tagout devices
- Lockout procedures
- Reenergizing procedures
- Knowledge of the equipment being serviced

Each *affected* responder (one whose job requires him or her to operate or use a machine on which servicing or maintenance is being performed, or whose job requires him or her to work in an area in which servicing or maintenance is being performed) will receive training on the energy control procedures that includes, but is not limited to:

- Immediate recognition of lockout/tagout devices
- The purpose of lockout/tagout devices
- How to determine when a control procedure is in use
- How not to disturb lockout/tagout devices
- Awareness of the danger of violating or disregarding procedures

All other responders whose work operations are or may be in an area where energy control procedures might be used will also be made aware of the procedures and the prohibition relating to attempts to restart or reenergize machines or equipment that are locked out or tagged out.

Limitations of Tags

All responders will also be trained in the following limitations of tags:

- Tags are essentially warning devices affixed to energy-isolating devices and do not provide the physical restraint on those devices that is provided by a lock.
- When a tag is attached to an energy isolating means, it is not to be removed without permission of the authorized person responsible for it, and it is never to be bypassed, ignored, or otherwise defeated.
- Tags must be legible and understandable by all authorized responders, affected responders, and all other responders whose work operations are or may be in the area in order to be effective.
- Tags and their means of attachment must be made of materials that can withstand the environmental conditions encountered in the workplace.
- Tags may evoke a false sense of security, and their meaning must be understood.
- Tags must be securely attached to energy-isolating devices so that they cannot be inadvertently or accidentally detached during use.

Responder retraining will be provided for all authorized and affected responders whenever there is a change in their job assignments; in machines, equipment, or processes that present a new hazard; or in the energy control procedures. Additional training will also be conducted whenever a periodic inspection reveals, or the safety officer has reason to believe, that there are deviations from or inadequacies in the responder's knowledge or use of these energy control procedures.

The safety officer will conduct a periodic inspection of the lockout/tagout program to ensure that the procedures and requirements of the standard are being followed. These audits will be conducted for each authorized responder at least annually. During these audits, responders will be evaluated individually, through random audits and visual observations, to ensure they can demonstrate how to correctly lockout equipment. These inspections will be documented, including who performed the inspection (someone other than the person performing the lockout/tagout), the identity of the machine or equipment on which the procedure is being used, the date of the inspection, and the responders included in the inspection.

HAZWOPER Training Requirements

This section discusses Hazard Waste Operations and Emergency Response (HAZWOPER) training requirements and levels of training commensurate with the responders' designated roles within the response operation. OSHA letters of interpretation (see "Resources" section at the end of this chapter) specify HAZWOPER first responder operations level training to first responders who are expected to perform decontamination duties or handle victims before they are thoroughly decontaminated. This level of training is appropriate for anyone with a designated role in the decontamination zone. Table 4.3 outlines first responder mandatory and recommended training.

Training requirements for the first responder operations level include a minimum training duration of eight hours and competencies the responder must acquire. Both the required competencies and training time are confirmed in an interpretive letter. OSHA, however, allows these topics (but not the minimum training time) to be tailored to meet the needs of first responders. For example, the training might omit topics that are not directly relevant to the responder's role (e.g., recognition of Department of Transportation placards), but instead should include alternative training on hazard recognition (e.g., signs and symptoms of contamination or exposure), on decontamination procedures, and on the selection and use of PPE. Training that is relevant to the required competencies counts toward the eight-hour requirement, even if the training is provided as a separate course. For example, training on PPE that will be used during victim decontamination activities may be applied toward the eight-hour minimum operations level training requirement, regardless of whether the PPE training is conducted as part of a specific HAZWOPER training course or as part of another training program.

First responder awareness level training also counts toward the eight-hour requirement for operations level training. This point is clarified in a letter of interpretation issued by OSHA: "If you spend two hours training employees in the required competencies for First Responder Awareness Level as described in 29 CFR 1910.120(q)(6)(i)(A)-(F), then you would need to spend at least six additional hours training employees in the required competencies for First Responder Operations Level as described in 29 CFR 1910.120(q)(6)(ii)(A)-(F). Depending on the employees' job duties and prior education and experience, more than eight hours of training may be needed."

As an alternative to the eight-hour training requirement, the HAZWOPER Standard allows employees to demonstrate competence in specific areas, presented in 29 CFR 1910.120(q)(6)(ii). OSHA reaffirmed this point in a letter of interpretation, which states "Employees with sufficient experience may objectively demonstrate the required competencies instead of completing eight hours of training." However, it is important to note that in most settings it might be difficult to ensure that employees have sufficient experience to waive the training requirement. Many responders may not have extensive experience with hazardous materials or PAPRs (powered air purifying respirators), and decontamination activities are performed infrequently.

Documentation is required regarding how training requirements are met. This is particularly important whenever responders are allowed to satisfy any portion of the training requirement through other related training or through demonstration

Table 4.3   Training for First Responders

Mandatory Training	First Responders Covered
First Responder Operations Level Initial training Annual refresher Both initial and refresher training may be satisfied by demonstration of competence.	All employees with designated roles in the Decontamination Zone. This group includes, but is not limited to, the following: • Decontamination staff, including decontamination victim inspectors, clinicians who will triage and/or stabilize victims prior to decontamination, security staff (e.g., crowd control and controlling access to the ED), set-up crew, and patient tracking clerks.
Briefing at the time of the incident	Other employees (e.g., a medical specialist or trade person, such as an electrician) whose role in the decontamination zone was not previously anticipated (i.e., who are called in incidentally).
First Responder Awareness Level Initial training Annual refresher Both initial and refresher training may be satisfied by demonstration of competence.	a. Security personnel, set-up crew, and patient tracking clerks assigned only to patient-receiving areas proximate to the decontamination zone where they might encounter, but are not expected to have contact with, contaminated victims, their belongings, equipment, or waste. b. Clinicians, clerks, triage staff, and other employees associated with emergency departments who might encounter self-referred contaminated victims (and their belongings, equipment, or waste) without receiving prior notification that such victims have been contaminated.

Continued

Table 4.3 (*Continued*) Training for First Responders

Recommended Training	Personnel Covered
Training similar to that outlined in the Hazard Communication Standard	Other personnel in the postdecontamination zone (e.g., other emergency department staff, such as housekeepers) who reasonably would not be expected to encounter or come into contact with unannounced contaminated victims, their belongings, equipment, or waste.

of competence. The HAZWOPER Standard requires and an OSHA letter of interpretation confirms that "the employer must certify in writing the comparable training or demonstrated competencies."

Annual refresher training is specified under 1910.120(q)(8)(i); however, the length of the refresher training is not specified. Instead, the standard requires that responders trained at the first responder operations level "shall receive annual refresher training of sufficient content and duration to maintain their competencies, or shall demonstrate competency in those areas at least yearly." Additionally, it is important to document that refresher training was performed or, alternatively, keep a record of how the responder demonstrated competency.

First responder awareness level training is required for those employees who work in the contaminant-free postdecontamination zone, but who might be in a position to identify a contaminated victim who arrived unannounced. This group includes clinicians, clerks, and triage staff who would be responsible for notifying authorities of the arrival, but would not reasonably be anticipated to have contact with the contaminated victims, their belongings, equipment, or waste. The group also includes decontamination system setup crew members and patient-tracking clerks, if their roles do not put them in contact with contaminated victims, their belongings, equipment, or waste (e.g., setting up the decontamination system before victims arrive or tracking patients from a location outside of the decontamination zone).

First responder awareness level training also is required for security personnel who work away from the decontamination zone, but who may be involved tangentially in a mass casualty event (specifically, those security personnel who would not reasonably be anticipated to come in contact with contaminated victims, their belongings, equipment, or waste). Security staff assigned to roles in the decontamination zone would require a higher level of training (e.g., first responder operations level).

Training requirements for first responder awareness level appear under 29 CFR 1910.120 (q)(6)(i), which does not require a specific minimum training duration but outlines topics to be covered (competencies the employee must acquire). As with operations level training, the HAZWOPER Standard allows an alternative to the awareness level training requirement. Training can be waived if the responder has had sufficient experience to objectively demonstrate competency in specific areas. These areas are listed in 29 CFR 1910.120(q)(6)(i).

Annual refresher training is required for responders trained at the awareness level. As with operations level refresher training, the class content must be adequate to maintain the responder's competence. Training or the method used to demonstrate the responder's competence must be documented.

Medical Protective Practices

To augment operational health and safety programs, medical protective practices should also be considered to protect responders during emergency operations. Medical monitoring can happen in the field as a direct requirement of health and safety programs and as on ongoing program to ensure that changes in condition are addressed through appropriate mitigation strategies. Pre- and postincident medical countermeasures may also be necessary to address responder health and safety, particularly during response scenarios that may expose responders to pathogens.

Medical Monitoring

It is important to determine that personnel who are being asked to wear PPE during response operations have no preexisting medical conditions that might put them at increased risk for illness or injury (Figure 4.7). As the response operation is beginning, medical monitoring equipment, such as blood pressure cuffs, stethoscopes, scales, thermometers, and medical monitoring forms, should be used by trained professionals to monitor all affected responders prior to beginning assigned tasks. If there is inadequate time to perform pre-entry medical monitoring, it is important that each staff member exercise good judgment and proceed with response operations only if they know there is no preexisting condition that should preclude their use of PPE. Clinical data obtained from medical monitoring must fall within appropriate medical guidelines. Persons whose vital signs exceed requirements should be allowed to rest for fifteen to thirty minutes and then be re-examined or given a responsibility not requiring the use of PPE.

Using the same or similar criteria used in preentry monitoring, responders should have postentry medical monitoring performed as well, and data should be recorded. Significant changes in clinical data may indicate the need for more comprehensive evaluation or medical treatment, and responders in this category should be placed out of service and evaluated further or treated.

After response operations are complete, medical monitoring records for all staff should be reviewed by a medical doctor (MD) or occupational health MD

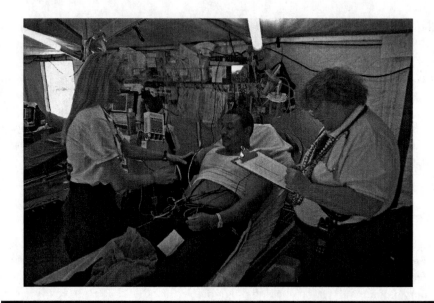

Figure 4.7   Personnel must be medically cleared to wear PPE. (Courtesy FEMA, Marvin Naumin.)

to determine if any further short- or long-term clinical evaluation is necessary. If the decision is made that additional evaluation is needed, the responder should be notified immediately and arrangements should be made for the exam. Responder records should be placed in their personnel file and kept secure and retrievable for a period not less than thirty years.

In many cases, mental health can also be impacted by response operations as well. Particularly in catastrophic incidents, it is important to pay attention to mental health and to offer critical incident stress management services as the need arises. In all cases, qualified medical and mental health professionals should be used in medical monitoring.

Vaccinations and Medical Countermeasures

If possible, vaccinations and medical countermeasures should be considered on a preincident basis. Using similar hazard assessment procedures discussed previously for PPE, it is possible to identify the most probable potential agents to which responders may be exposed. For example, the Michigan Department of Community Health has developed the recommendations for providing vaccinations to emergency responders as seen in Table 4.4.

Biological and pathogenic risks should be assessed on a local, regional, and functional basis to determine the most appropriate vaccines and medical countermeasures to employ. In addition to vaccines, antivirals and other medications may provide utility on a case-by-case basis.

Table 4.4   Recommended Vaccinations

Vaccine	Recommendation in Brief
Hepatitis B	Give three-dose series (dose #1 now, #2 in one month, #3 approximately five months after #2). Give IM. Obtain anti-HBs serologic testing one to two months after dose #3.
Influenza	Give one dose of TIV or LAIV annually. Give TIV intramuscularly or LAIV intranasally.
MMR	For healthcare personnel (HCP) born in 1957 or later without serologic evidence of immunity or prior vaccination, give two doses of MMR, four weeks apart. For HCP born prior to 1957, see below. Give SC.
Varicella (chickenpox)	For HCP who have no serologic proof of immunity, prior vaccination, or history of varicella disease, give two doses of varicella vaccine, four weeks apart. Give SC.
Tetanus, diphtheria, pertussis	Give all HCP a Td booster dose every ten years, following the completion of the primary three-dose series. Give a one-time dose of Tdap to all HCP younger than age 65 years with direct patient contact. Give IM.
Meningococcal	Give one dose to microbiologists who are routinely exposed to isolates of *N. meningitis*.

Priority Prophylaxis

Responders may be deployed unknowingly to an event that could result in exposure to acts of biological terrorism or outbreaks of infectious diseases. Thus, it is important to have pre-positioned regimens of prophylactic medications available for responders and their families. As with the strategic national stockpile (SNS) of medications, the most common prophylactic (preventative) medications are amoxycillin, ciprofloxacine, and doxycycline. Many states and local jurisdictions have already developed priority prophylaxis programs to dispense necessary medications to first responders and their immediate household members as quickly as possible once a threat has been identified.

Such programs require procedures for notification of first responders, predetermined numbers of responders and immediate family members, and predetermined locations for dispensing. Priority prophylaxis programs ensure that essential first responder personnel receive prophylaxis in a timely manner when responding to a bioterrorism event or outbreak of infectious disease. It is important that responders are able to deploy to assigned duties with confidence that they and their families are protected. This peace of mind will allow them to effectively perform their assigned responsibilities during a time of critical need.

Once priority prophylaxis is performed, responders can begin setting up mass dispensing sites while awaiting the arrival of the SNS or other drug caches. Upon receipt of SNS supplies and medication and prior to the opening of the mass dispensing sites to the general public, other SNS team members, volunteers supporting SNS functions, and their immediate household members can receive prophylaxis from the SNS. This model may also be effective in distributing antivirals and providing vaccinations, particularly if new strains require different vaccines or treatment.

Conclusion

Personnel responsible for saving lives and keeping the public safe must keep themselves safe. Employees may be more aware of their standard workplace safety measures, but they must also be prepared for emergency situations, which can be confusing and volatile. Constant training and documentation are crucial for responder safety, and there are many organizations, governmental and otherwise, that create and update not only guidelines, but also rules, to ensure responder safety. A responder's most valuable safety assets are preplanning and prevention.

Resources

American Conference of Governmental Industrial Hygienists: www.acgih.org
American National Standards Institute: www.ansi.org/
National Fire Protection Agency: www.nfpa.org/
National Institute for Occupational Safety and Health: www.cdc.gov/niosh/
Occupational Safety and Health Administration: www.osha.gov/
Strategic National Stockpile: http://emergency.cdc.gov/stockpile/

Chapter 5

Disaster Preparedness and the Law

Thomas Schneid

Contents

> The minute you read something you can't understand, you can almost be sure it was drawn up by a lawyer.
>
> **—Will Rogers**

The Questions

Why should your city or county attorney or any attorney be included on your emergency and disaster preparedness team? What does a lawyer know about emergency planning? What does an attorney bring to the table in the emergency planning process? Will the attorney add value to the team or be a detriment to the overall planning process?

When Dr. Fagel asked me to address this issue, the first and probably the best reason that any emergency and disaster preparedness team would want an attorney as a member is because of the axiom, "No good deed goes unpunished." (Quote has been attributed to a number of different authors including Oscar Wilde, Joe Orton, Isadore Pavia, Andrew W. Mellon, and Clare Booth Luce.) Think about that quote as you read this chapter and consider the points herein.

An emergency and disaster preparedness team is assembled based upon individuals' knowledge, skills, and abilities, their expertise, and their experience with the primary goal of the team being to "do good deeds" for others in an emergency or disaster situation. Isn't it logical and prudent to have at least one member of an emergency and disaster preparedness team with a skill set that is particularly focused on the legal aspects of a team's actions and possible inactions that could inadvertently cause harm to others?

I have to admit, I am probably biased toward the inclusion of an attorney on emergency and disaster preparedness teams. You see, I am an attorney as well as a professor in the department of safety, security and emergency management program at Eastern Kentucky University. I'm not trying to squeeze in a chair at the emergency and disaster preparedness table for attorneys, but I would like to make the case that there is a definite need for an attorney's unique legal perspective as part of any emergency and disaster preparedness team especially from the beginning of the planning process. Please permit me to try to persuade you and to make my case for the inclusion of legal representation on your emergency and disaster preparedness team.

The Answers

First, we live in a very litigious society. The good deeds that each and every member of the team wants to do to assist the community during an emergency or disaster situation may result in some form of legal action, claim, or litigation directly from their good deed action or even possibly from lack of action, depending on the circumstances. An attorney on the emergency and disaster preparedness team can possibly identify potential pitfalls or potential legal risks wherein the emergency and disaster preparedness team can take proactive steps to protect against the risk, educate their members as to the risk, or even ensure against the risk to eliminate or minimize the risk from becoming a "disaster after the disaster."

Second, an attorney on the emergency and disaster preparedness team can bring a different perspective due to the specialized skill set that was drilled into them during their legal studies. Every emergency and disaster preparedness team, no matter how well prepared, is second-guessed following any event. A skilled attorney can assist the team in analyzing the potential risks during the planning process to provide proactive alternatives to eliminate or minimize the risks. Just as there are two sides to every argument, there are two sides to every situation, depending on your perspective. A skilled attorney should be able to provide the opposing viewpoint for the emergency and disaster preparedness team, especially during the planning stage, to permit the team to carefully analyze the situation and determine if the risk is acceptable or offer alternatives to counteract the risk.

Third, you definitely need a trusted advisor who will shoot straight with you. Although we may not recognize the hierarchy within our team or organization, some members may not wish to rock the boat when it comes to the decision-making process. What looks like a good decision today may ultimately turn out to be a very bad decision in the long run. Your trusted advisor will have your back and be able to offer advice and counsel that may avoid future difficulties.

Fourth, the intense training and education provided to attorneys in law schools often create a tendency to view issues in a very meticulous manner. A team is often assembled according to experience, expertise and personality, and each member is expert in his or her areas of expertise. How many of your team members are great at paperwork? Who on your team will double check the team's paperwork and check all of the boxes to ensure the accuracy and correctness of documents? Your attorney can serve as your fail-safe through which all documents can be evaluated to ensure that the team doesn't get egg on its face as a result of paperwork errors.

Fifth, just as the quote by Will Rogers at the beginning of this chapter noted, many of the grant proposals and other documents that may come before the emergency and disaster preparedness team may be written in legalese. The skilled attorney can interpret the legalese into a comprehensible verbiage for the emergency and disaster preparedness team members. This should permit the individual team member to focus on his or her areas of expertise when developing the grant or other document rather than spending time with a legal dictionary.

Sixth, the media is omnipresent in any emergency and disaster situation. Although many emergency and disaster preparedness teams possess an information officer, the skilled attorney can assist in honing any statements to the press and keep the proverbial foot from entering the mouth of team members. As we are all aware, video footage from any emergency or disaster situation locks the scene in time and is often utilized to Monday-morning quarterback the actions or inactions of the team in public forums as well as in litigation in the future.

Last, what is the downside of having an attorney on an emergency and disaster preparedness team? Another chair at the planning table? Another set of eyes and ears to add a new and different perspective to the planning process? Long story short, there is very little downside to adding a city attorney or local attorney to the planning process, and the substantial upside is the ability to proactively identify potential legal pitfalls during the planning process.

So, now that I have at least piqued your interest in possibly adding a city attorney or in-house/outside counsel to your emergency and disaster preparedness team, what potential legal liabilities or pitfalls should your team look for in the planning process that you can discuss with an attorney? The areas of potential liabilities in any emergency and disaster situation run the spectrum from workers' compensation to vehicle accidents to injuries occurring after the emergency and disaster team leaves the scene. And additionally, the applicability of the law depends on whether your emergency and disaster preparedness team is encompassed within the area of federal or state governmental immunity or other qualified immunity laws.

The Potentially Applicable Laws

Let's take a look at a few of the potentially applicable civil laws as well as the areas of potential liability or pitfalls in an emergency and disaster situation.

Federal laws—Including but not limited to:
- Americans with Disabilities Act (ADA)
- Occupational Safety and Health Act
- Labor Management Relations Act
- Federal Service Labor–Management Relation Act
- Civil Rights Act of 1964 and 1991
- Age Discrimination in Employment Act
- Veterans' Reemployment Rights Act
- Rehabilitation Act of 1973
- Federal Anti-Discrimination Executive Orders
- Fair Labor Standards Act (FLSA)
- Minimum Wage Requirements under the FLSA
- Overtime Pay under the FLSA
- Employee Retirement Income Security Act (ERISA)
- Federal unemployment laws
- Federal privacy laws
- Family and Medical Leave Act (FMLA)

State laws—Including but not limited to:
- State labor relations laws
- Anti-discrimination laws
- Protection for military personnel
- State wage and hour laws
- State safety and health laws
- State workers' compensation laws
- Governmental immunity statutes

Local laws—Including but not limited to:
- – Building codes
- – Jurisdictional requirements
- – Land use laws
- – Traffic laws

Although these are just a few of the myriad federal, state, and local laws and requirements possessing potential civil liability, there is a completely different set of laws on the criminal side. Generally, the criminal laws are applicable to individual team members' actions or inactions, rather than the overall organization. However, many of the aforementioned laws do possess a criminal component for egregious situations.

Negligence

One specific area in which an attorney can educate an emergency and disaster preparedness team is in the frequent civil actions sounding in the area of negligence. There are numerous causes of action in many areas sounding in negligence ranging from negligence hiring to negligence supervision to negligent injury.

Generally, to prove negligence, the injured party must provide:

1. Duty—The emergency and disaster team possessed a duty to act.
2. Breach—The emergency and disaster preparedness team breached this duty through actions or inactions.
3. Causation—The actions or inactions of the emergency and disaster preparedness team caused the injury to the other party.
4. Damages—The actions or inactions of the emergency and disaster preparedness team caused damages to the other party.

Although this breakdown is provided in a simplistic manner, negligence actions are common in federal and state civil courts against public and private sector entities. For public sector emergency and disaster preparedness organizations, there is the additional issue of immunity that would need to be addressed. An attorney, as a valued member of your emergency and disaster preparedness team, can educate the team as to the requirement of these laws in order to avoid risks and areas of potential liability as well as being the well-prepared advocate in case a legal action should arise against the organization or team.

Emergency Management Phases

Let's take a look at the various stages of an emergency and disaster preparedness plan to ascertain some of the general areas where potential legal issues could arise, in which legal counsel could assist an emergency and disaster prepared-ness team.

Planning Stage

- There are jurisdictional issues between state and local governmental entities.
  - Who is responsible for what activities?
- A mutual aid agreement is agreed upon between departments. One entity fails to perform the required responsibilities.
  - Who is responsible for any consequences that arise from this?

Response Stage

- The 9-1-1 system is not working.
  - Who is responsible?
  - Is the failure of the 9-1-1 system the cause of a grandma dying?
- A member of your emergency and disaster team renders first aid at the scene. The victim dies, and the family blames the first responder.
  - Who may have liability?

Recovery Stage

- The team is attempting to repair a gas line and the line explodes.
  - What if a team member is injured or killed?
  - What if the neighboring house is severely damaged?
- The team member is utilizing a chain saw and the chain saw malfunctions causing severe injury.
  - Is the manufacturer responsible? Your organization?
  - Is third-party recovery possible?

Other Areas of Potential Civil Liability Risks

Vehicles

- A member of your team is injured while driving a vehicle during an emergency situation.
  - Who pays the medical bills?
  - Disability for those involved both on your team and otherwise?
- A family brings a legal action for a team member killed in a vehicular acci-dent going to the disaster scene.
  - Can the family sue?

- A team member fails to stop at an intersection on the way to the scene and kills the other driver.
    - Can the family of the other driver sue?
    - Who would they name in the action?
- The brakes on the team member's official vehicle fail, and the vehicle injures a number of children when it crashes into a pizza parlor.
    - Can the children sue or will their parents sue on their behalf?
    - Who would they sue?

Employment

- A qualified 42-year-old employment candidate applies for a job with your team and is not selected.
    - Was the candidate discriminated against due to her age?
    - Is the candidate protected under the Age Discrimination in Employment Act?
- Your team adopts a "no beard" policy for safety reasons.
    - Did your team discriminate against your bearded team members?
    - Can the wearing of a beard be for religious reasons?
    - Does the wearing of a beard create a safety hazard?
- A team member is not promoted due to statements he or she has made in public with regard to a controversial animal rights issue.
    - Was the team member discriminated against due to his or her public statements or stance?
    - Are the team member's public statements considered free speech under the U.S. Constitution?

Discrimination

- A male team member continually asks a female member of your team out on a date which she refuses.
    - Is the male team member harassing the female team member?
    - Is it actionable under Title VII of the Civil Rights Act?
    - Would the action be brought against the team member or the team/department?
- The team leader promotes a white male team member with five years experience over a Hispanic team member with ten years experience.
    - Did the team leader discriminate against the Hispanic team member?
    - Is this actionable under Title VII?
- A candidate for your team is in a wheelchair due to a permanent injury. The team decides that the candidate was not the most qualified for the position.

- Did the team discriminate against the candidate under the Americans with Disabilities Act?

Training

- One of your team members has a heart attack and dies while training.
  - Are death benefits paid under workers' compensation?
  - Can the family bring a legal action?
- A member of the team has a cold and refuses to participate in the training exercise. The team leader dismisses the team member from the team.
  - Can the team member bring a legal action against the team?
- A team member drops a pipe on the foot of a fellow team member during a training exercise.
- Can the injured team member sue the team member who dropped the pipe?

These are but a few situations where an attorney can be of assistance in providing guidance to the team, but these are only a few of the situations that can arise as a result of the myriad federal, state and local laws. And these are on the civil side of the aisle. There is an entirely different set of federal, state, and local criminal laws that can impact the team or any team member.

Summary

A professional with specialized legal training can be a beneficial member of an emergency and disaster preparedness team. We are all aware of the zillions of jokes about lawyers and the high opinions many possess about the legal profession (right?); however, lawyers do bring a lot of experience and specialized expertise to the table. Every emergency and disaster preparedness team will encounter legal questions and situations where the appropriate legal advice can save time, money, and even lives.

Remember, the legal professional is now part and parcel of your team! The legal professional will learn and grow as part of your team, and I'm sure he or she will become an indispensable team member in short order. What have you got to lose? And more important, what will your team gain?

The attorney on your team becomes your attorney and can research your situation and provide guidance through which you avoid situations where potential legal liabilities may exist before, during, or after an incident. He or she provides you guidance in order to minimize your potential liability.

Isn't it nice to have your own in-house attorney as part of your emergency and disaster preparedness team?

Chapter 6

Stress Management and Responders

K. R. Juzwin

Contents

Why Stress Management Is Important to Emergency Managers

In crisis situations, community management has two populations to attend to and manage: their civilian citizens and their own employees. This chapter offers emergency planners and community management professionals a framework of important components to build into their emergency response plan in a predisaster situation. There are a number of benefits for building this structure. The first is obvious: having a solid plan ready to put into place in a disaster or emergent situation helps optimize success in response and recovery. But second, if the organization is focused on the long term, consideration of specific elements has been shown to predict organization commitment. Organization commitment has been shown to directly relate to the perceived leadership style, especially where the leadership defines roles of personnel toward defined goals through assigned tasks, using specific procedures, timelines, and communication (Dale and Fox 2008). Further, research demonstrates that planning targeting reduction of perceived role stress and increasing or solidifying interpersonal factors have benefits to the organization (Dale and Fox 2008).

Ever evolving, emergency managers have begun to anticipate events that meet and exceed the scope of what they can manage within their own municipality, region, or state. The mechanics of the response structure appears well dictated, as evidenced by the trend to use the National Incident Management System (NIMS) approach, and the protocol of various agencies mandating response within their scope of service. In the case of disasters, what happens? "On rare occasions, emergencies occur that are so large in scale and so severe that local responders may not have the resources—people, equipment, expertise, funds—to effectively and safely respond. Even in such cases, local responders do not hesitate to do what they have been trained to do—go to the site prepared to save lives, protect property, and remove the threat" (Jackson et al. 2004, iii).

Law enforcement has built into its preincident strategic planning interventions to add to its incident command approaches. Five components have been identified as necessary to build into preincident planning (Sheehan et al. 2004). This includes assessment and triage, specific crisis intervention with individuals, and to small and the large group crisis intervention, and strategic planning. Within the context of an emergency situation, the community manager will want to optimize the chance for retention and ability to return to function in a typical daily work routine. Strategic planning should take into account elements of the design of tasks, management style, interpersonal relationships, work roles, career concerns, and environmental concerns (Goetzel et al. 1998). Specifically, each of these are important to consider in their contribution to stress for the employee:

- The design of tasks involves heavy workload, infrequent breaks, long work hours, shifts, tasks with little personal meaning, underutilizing of skills, and lack of control
- Management style where employees do not contribute in decision making, poor communication, and absence of family-friendly policies
- Poor social environment and interpersonal relationships among coworkers and supervisors

- Work roles that have conflicting or uncertain job expectations, or too much responsibility
- Job insecurity, lack of potential for advancement, or rapid changes for which workers perceive they are unprepared contributes to career concerns
- Unpleasant or dangerous physical conditions contribute to environmental concerns

Minimal attention in emergency planning has focused on the potential secondary survivors, specifically the responders and emergency service workers who attend to all of those demands. One element often not accounted for in planning is the nonemergent and nonimmediate human impact element for the responders. It is important to consider the well-being of the employees because when a disaster strikes in their community, they become victim-responders. Their commitment will be to do what they have to do for their job and community. Their minds and hearts will be preoccupied with their own families, homes and neighborhoods. It is important to note that there are findings (Galea 2007) suggesting that the serious long-term consequences of disasters and mass trauma are most frequent in specific groups of survivors. In order of severity of consequences, the people injured in the incident, rescuers, people who have lost their personal belongings and homes, families of those injured, and then the general population (Galea 2007; Sheehan et al. 2004).

Stress-Related Disorders

It is estimated that stress-related disorders will be the second leading cause of disability by the year 2020, and is the primary strategic goal for the World Health Organization's Global Burden of Disease of the NIOSH Work Organization and Stress-Related Disorders Program (Ray and Sauter 2008). In the United States, work-related stress is estimated to cost $171 billion annually, which is the same as cancer and cardiovascular disease, and greater than Alzheimer's or HIV or AIDS. There is an estimated $300 billion annual cost due to lost hours from absenteeism, decreased work productivity, and cost of health expenditures (APA 2004).

Organizations such as the World Health Organization and the National Institute of Mental Health suggest that emotional health-related problems, suicide, and mental illness account for over 15 percent of the burden of disease costs to an economy. The number one cause, cardiovascular conditions, contributes to more than 18 percent of the disabling conditions. However when alcohol and drug use problems are factored into the emotional-problem-based disabling conditions, this increases disability to more than 20 percent.

In a disaster situation, the rate of impact is generally greater. Research demonstrates that individuals without the benefit of prior disaster training or experience are at greater risk for posttraumatic stress disorder (PTSD; Perrin et al. 2007).

In a study of PTSD post-World Trade Center, the rate was 21.2 percent for construction, engineering, sanitation and unaffiliated workers, compared to a rate of 12.4 percent for rescue/recovery workers (Perrin et al. 2007). The National Mental Health Information Center (DeWolf 2005), there are several important considerations regarding people who have any exposure to disaster situations. While all of these are important, one is absolutely the most critical point. These points show the importance of highlighting why it is important to consider the mental health needs of not only primary survivors, but also of the secondary survivors and response personnel. There is absolutely one most critical point: No one who sees a disaster is untouched by it.

Understanding Stress along the Continuum

Stress occurs as a function of living. Stress reactions are normal, but can be adaptive or maladaptive. Stress can be motivating and positive, or negative and destructive. In extreme situations, reactions can be viewed as "normal responses to abnormal events rather than signs of psychopathology" (Rogers 2007, 3). Stress can arise in the short-term, and in response to immediate demands, and is reasonable to expect as the situation intensifies and grows. No one is immune from stress, but building a skill set, an expectancy mindset, and developing a response plan can help avoid some of the destructive effects of it. Typical stress symptoms can impact the body's ability to be healthy and can interfere with the ability to fight illness or infection. Stress, at its extreme, is connected to heart disease, cancer, respiratory problems, accidents, suicide, depression, anxiety disorders, and alcohol-related problems. As discussed earlier, these contribute to the largest debilitating illnesses and economic impact. This impact has direct and indirect costs to the individual, employer, and society. These costs include days out sick, paid/unpaid time off, shift or workload coverage, treatment, recovery, and lost productivity to the employer.

When something happens that is out of the norm, the potential for problematic stress occurs. This event can be a storm that is far worse than usual, a damaging tornado, a significant fire or flood, or other type of event that has far reaching impact in regard to length of time, extent of damage, extent of response needed, and recovery from the event. Exposure to these situations causes reactions that may include terror, horror, and fear for one's safety and well-being. Reactions may be anywhere from mild to severe, and impact people in a number of different ways over a long time. Most reactions are temporary, although these can be quite unnerving to the individual because they are so out of the range of typical reactions that usually happen for the person. Symptoms are patterns of behaviors, reactions, feelings, thoughts, or problems that impact how one manages in his or her life.

First responders and support personnel in emergency situations performing rescue and recovery work are exposed to stressors beyond their normal daily emergency-task demands. Their "normal" emergency response is generally above and beyond the situations that most of the general population ever faces. However, when there is a situation outside of that "normal" or "typical" scope, exposure to these events increases the risk of emotional and physical trauma, and PTSD. Whereas the general population is considered to be at risk for PTSD at a prevalence rate of about 4 percent, this increases upward of 32 percent for rescue and recovery service personnel in the World Trade Center study, 25 percent for search and rescue, and 21 percent for firefighters (Perrin et al. 2007).

Stress Reactions

There are a number of symptoms that interfere with emotions or feelings, physical health or with thinking. Because these problems affect our thinking, feeling, and physical functioning, our behaviors and the way we manage in our daily lives can become problematic and influenced in many ways.

Most of us have a typical baseline for recovering from stress and stressful events in our lives. We may be thrown off for a period of time, for example, a couple of days. However, when these problems are not resolved, then stress may be moving to distress or even cumulative stress. At the worst, these cause a negative impact on physical health as well.

It is important to note that everyone handles and responds to stress differently. Some people have immediate signs of stress; others will not demonstrate any. Sometimes it is very obvious, as some of the symptoms show dramatically in physical complaints (see Table 6.1). Other people show only subtle signs that can often be attributed to something else entirely. When these signs last for more than a few days, it can be a signal that the problems may be more than the typical stressor, and that maybe some additional support might be needed.

Educating people on stress in terms of physical, emotional, or cognitive signs gives them an opportunity to watch for changes in themselves and in each other. Creating a culture of positive self-care and encouraging people to practice stress management daily is a solid foundation for preventative care for future situations.

Acute Stress

Acute stress reactions are generally very distressing to the individual, and often the individual doesn't recognize these reactions as connected to the event. And in emergency situations, everyone can be so consumed with responding to the crisis that people may not realize that either they themselves or their peers are experiencing acute stress response. In the heightened focus of the moment, people may not be aware of their own reactions. There may be panic, shock, disbelief, disorganization, or distress. Many people experience a sense of being out of control.

One experience is described as feeling "disconnected" or feelings of disorientation or that the individual is "going crazy." Unfortunately, people often also feel as though they are losing control or unable to manage, which can be compounded when, in fact, their ability to manage is somewhat compromised. It is important to make certain to connect the problems as reactions to the event, and that when managed, they tend to be lessened.

People who experience this acute stress report a variety of problems and changes in their behavior. These changes or problems occur in feelings, thinking, physical functioning, and behaviors. These problems include those listed in Table 6.1, and can be very intense and disturbing to the individual. *This is especially problematic if they are in the form of recurrent images, smells, sounds, or thoughts about the stressful events.*

Manager Responsibilities in a Critical Event

There are two levels of attention that managers must take into account during a critical event. First is their responsibility in organizing and implementing their crisis response plan. The second is their responsibility to their people responding to and implementing the crisis response plan. Without the second piece being managed effectively, the first piece may be compromised. Each person involved in responding to a community plan will have families, loved ones, friends, and property, and worries about each of them. While they may be the emergency response team, they are also human beings with human connections.

Understanding stress responses helps insulate people and normalize stress. It also helps them recognize that when stress symptoms become too much, it is important they are encouraged and supported to get the help they need before it becomes traumatic or critical incident stress.

Planning: Helping Take Care of Your Responders in Advance

The cornerstone of public health includes accessing local to international resources to assure health of the people. Preplanning and having all of these elements in place may help alleviate some aspects of stress for employees, knowing they have provided the best for their family as they were able, even in their absence from the family.

Policy dictates and guides our eventual practice, like a blueprint dictates the building of a house. Many administrators understand and accept the need for moving beyond the reactive. Traumatic stress experts advocate that there should be representation at the policy making level, to offer guidance, consultation and ongoing collaboration (Fairbank and Gerrity 2007). Using an approach that includes

Table 6.1   Common Symptoms of Stress

Physical Symptoms	Emotional Symptoms	Cognitive Problems
Constant fatigue or lack of energy	Feeling irritability or short tempered	Feeling overwhelmed or unable to manage demands in daily life
Inability to relax	Sad, blue, or down	Problems with remembering things and new learning
Sexual problems	Worry, apprehension	Forgetfulness
Weight changes	Isolating self, feeling isolated	Problems with problem solving
Cold sweaty palms	Crying, tearfulness	Difficulties concentrating
Changes in sleep patterns, trouble falling asleep, poor sleep or lack of restorative sleep	Lessened ability to relax or to enjoy things; feeling abandoned or targeted	Difficulties making sense of things; slowed thinking
Difficulties with attention or poor concentration	Feeling out of it, overwhelmed, ineffective	Rigid or obsessive thinking
Aches, pains, and muscle tension	Thinking of death as a an option; hopelessness	Problems with word finding
Headaches	Anxious, fearful	Reliving an event over and over
Grinding teeth, clenched jaw	Panic, terror, hypervigilance	Denial or blame
Stomach and gastrointestinal problems such as upset stomach, nausea, diarrhea, and indigestion	Compulsions, checking, putting things in order; anger/rage	Difficult dreams; denial/blaming; suspiciousness
Loss of appetite	Apathy, lethargy; no energy	Doubting faith
Racing heart beat, shallow breathing, tightness in the chest	Numbness	disorientation, confusion
Increased startle response; inability to relax	Behavioral outbursts	Easily overwhelmed

an emphasis on psychological health; early identification of psychological problems followed by early treatment and intervention; and cultural relevance, and elements involving family and community support is strongly recommended to be considered in public health policy (Fairbank and Gerrity 2007).

There is at least one way to help mitigate the stress incurred by responders in a disaster situation: *assure their families are as safe as possible because of anticipated preparedness.* The family is prepared to respond, and therefore in a more secure position to react and be self-sufficient. Controlling those elements that can be foreseen and premanaged is the foundation to any good strategic plan.

In the development of the community response plan, it is necessary to emphasize the importance that all employees have emergency plans in their own homes for their families that would include all the necessary general elements that are recommended for all citizens. It is a general rule of thumb that citizens should plan to be able to be self-sufficient for up to three to four days (72 to 96 hours). If the disaster is of severe enough scope, many elements will not be able to be supplied for a longer period of time.

What elements go into a family emergency plan? In these times, almost every village or city has a Web site that directs its citizens to an emergency plan. One of the most thorough is presented by the Federal Emergency Management Agency (FEMA) at www.fema.gov.

A general family plan should at least have elements including:

■ Emergency planning for potential risks and specific hazards
■ Emergency financial planning
■ Emergency and alternative communication plan and contact numbers, especially if in the event of mass communication disruption
■ Designated contact and/or meeting places and contact schedule
■ Disaster supplies kit for each family member
■ Disaster supplies kit for each pet
■ Alternative shelter arrangements
■ Food and water
■ Sanitation needs
■ Cash monies and credit cards
■ Plan for communication, heat, light, power, and so forth
■ Anticipating the environmental risks and the elements, including bugs/wildlife
■ Protective clothing, gloves, and hats

In addition to these components, the agency can provide suggestions to encourage employees to have their crisis response plan in place, so if they aren't there to assure the safety of their families, their family members know what to do related to:

■ Where to meet in the house, community, or out of the danger area
■ Who to contact about status, especially if separated

- Important documents and money, access to it if necessary
- Emergency supplies, resources, documentation, and identification (emergency preparedness lists)
- Resources within the community for contact
- Household information (gas, electric, etc.)
- Identification procedures and documentation
- What the rules are, and that everyone knows the plan
- Who is responsible for who and what during an evacuation
- Who takes what with them and what gets left behind
- When to evacuate
- How to communicate the plan, with written information of contacts, numbers, backup plan, identified people everyone will contact
- Who to go to for help if necessary
  - A call-tree for resources and supports
- When to leave and where to regroup and other backup plans
- Identification, insurance, and vital records, inventory of home possessions
- Information related to utility shut-off and safety

Here are some ideas to build into the emergency plan the following elements that focus specifically on anticipating the needs of employees for their families and homes.

- Educate all employees about the specific agency policies and expectations for their responses during a major incident or disaster. By informing them in advance that the scope of their jobs has demands of community response, they can anticipate the conflicts that arise between the professional and personal demands. If the agency has a specific plan, put it in succinct relevant information in bullet point formats, date it (so there is always a time reference), and distribute it. The goal is that this information can be put in the employee's home disaster kit.
  - Encourage employees to have a disaster or emergency plan at home and to talk about it in advance of any situation.
- Include an avenue for communication and contact for employees and their families. This can include an identified go-between who has the responsibility of being the contact person and relaying of information in a planned and anticipated manner.
  - Assure the agency has current updated emergency contact information and an identified person for each employee. This should be portable and taken with the communications officer. It should be in electronic as well as hard copy.
  - Identify a point of contact for the organization and how that contact should be made for the employee and for the family who may be trying to contact the organization.

- Develop a system for communication of status that involves a backup plan for this information to be conveyed in the event of local communication failure.

■ Have a system in place to assure that your responders can have communication with their family members and support persons. Build in a mechanism, time, and a place for employees to have status contact with their family contact person. If they have a sense they know their families are safe, they can focus on what their job is with more clarity and less stress. There may be less walking off the job or quitting if responders are able to have information about the status of their first priority, their families.

■ Encourage employees to have as part of their own disaster response strategy an ability to reach out and connect with nondeployed employees or responders to be available to support families of deployed responders to assist them to assure the families' welfare and safety in their absence. If this is not possible, have the employees identify other neighbors or friends who can help out. Everyone should have someone they are attached to and cooperating with, to get them through the immediate crisis and danger phases, and into recovery phase. This is particularly important for prolonged service and absence.

- Encourage employees to develop a network of spouses/significant others and families who can provide support, relief, and help to the family during a deployment. Include a call tree or flow chart of who can provide what resource to others. Many spouses of military have regular support group meetings and functions to offer one another support and resources during the deployment of their spouses.

■ Identify resources within the community that would be available to assist families in the event of an emergency. This may include church organizations, other coworkers, service agencies, schools, and so forth. Your plan might include developing support agreements among your community agency and other community resources so that they know to extend their support to these families. Additionally, encourage your responders to develop relationships within their community so there are established connections before something happens. Identify who is available within your community to act as a center for various supports (family reunification, child services, or homeless).

■ Anticipate a centralized and accessible for responders only, a place where information can be posted. This can include a schedule of informational briefings, centralized posting of information, and posted schedules of when updates are going to occur.

Suggestions for Supporting Your Responders

There are several strategies to assist your responders to organize themselves and attain their goals while lessening their stress levels during a critical event.

■ Communication is key. Have regular planned briefings with your personnel. This helps people anticipate when information is going to be given. The briefing should occur at intervals that are known and anticipated. Make it easy for your people to know who is who, and who to go to for what.

■ When appropriate, post information in a centralized location for those who are not able to be at a briefing but need the information. Make certain there is nothing in a posted brief that you would fear a reporter getting a hold of and put posted briefs in secure locations.

■ Make certain that during the briefings your lines of communication are never cut, put on hold, or sent to phone mail. Always be able to have your command center accessible by a live person who can interact with the caller.

■ You may want to have a meeting at the end of a shift or when people go off duty, to thank them for their efforts. When possible, use their names, shake hands, pat them on the back and make eye contact. Ask about their family's well-being during this time. Tell them their efforts are appreciated and noted.

■ Allow for clear lines of communication and responsibility. People forget to communicate the information about the whys and the specifics of why things are and are not happening. Your plan might want to post this information, to demonstrate that requests for resources have been acknowledged and that there is legitimate reason for why a supply may not be here. Status reports are vital; they help people anticipate and gain a sense of control. Include anticipated times, dates, and so forth when you know them. When you don't have that information, indicate that information is, as yet, unavailable but will be provided when received.

■ Have a place for your responders and workers to go to escape the situation, and rest and refuel. This may mean assigning someone to make sure that there is available nourishment available all the time. This can also be a role for a mental health clinician or chaplain who can be available for status check-ins for the workers.

  – For example, as a devastating crisis was unfolding, one hospital put out sandwiches, cookies, coffee, water, and cold drinks. People would gravitate there just to rest. People who worked there but were off shift showed up just to be there with their coworkers. The administration blocked the media from the grounds, and escorted people to and from their cars. Hardly anyone spoke, a number of them sat there crying, but the unspoken support was there. The crisis responders walked around the cafeteria, offering support, contact, and, when asked, a prayer. As the days postincident passed, the hospital chaplain and crisis workers continued to be present and prayed with some people and sat and talked about anything with people. The loyalty to the agency increased, and the organization's morale pulled them through this difficult time with few people choosing to leave their jobs because of the incident.

■ Your command staff should watch for signs of stress as discussed earlier. Irritability, loss of temper, anger, withdrawal, and hostility are big cues that your staff is stressed and potentially struggling. Remind them to give people time down for a quick break, and if they won't take it, ask them to take it anyway. Acknowledge their desire to stay focused, busy, and involved, but not at the expense of their health or burn out. When someone loses their temper or their cool, give them time to regroup. They've hit their wall, and need a minute or two to get it back together.

■ It is important to monitor if any responders are becoming angry, hostile, irritable, or acting out of their normal range of typical responses. For managers, it is important to see this as important information about their stress tolerance, fatigue, or overload. Often emergency workers can't or won't say they are scared, overwhelmed, and so forth. Many times, out of fatigue and exhaustion, they can't recognize it in themselves. If they had to repeat their experiences to others of what they see or do, they wouldn't be able to be out there doing what they do. So, they've developed ways of managing the awful stuff that overwhelms the everyday person. It is important not to personalize their behavior or reactions but to recognize it people stretched even beyond their own limits. Ask them what they might need or want. It may be reasonable; it may not be possible. Talk in private, give them some time to gather themselves and regroup. Don't retaliate or respond back in anger. Give them a couple of minutes have some quiet time and regroup. Check in with them later. They may say nothing; they may have a lot to say.

■ Watch for responders' reactions to one another. Make sure you don't have a scapegoat situation developing, where one person is the identified target for the groups' feelings of anger, helplessness, or being overwhelmed. Encourage their teamwork; effort as a team is what makes them the strongest they can be. If there is a personal overlap with any of them in this situation, watch for that. Your young team members may be very vulnerable. This is a time when you want to be aware of the group think; the team bond around the stronger members and cull out the one that is emotionally the one not like the rest. This group think behavior can cause the best of teams to crash and not be able to do what they've always relied on one another as a team to do.

■ Make certain your people have access to a routine and schedule. Routine establishes some element of predictability during times of uncertainty. Establish active and down routines.

■ Have a plan for removing someone if necessary and getting them the assistance they need.

■ Integrate support into maintaining team members' ability to keep doing their jobs, as a preventative and as a way for them to refuel and recharge, and reconnect with their teams.
  - Make sure everyone has known time in and time out of service.
  - When they are out of service, they need to be out of proximity of the situation as much as is possible. Additionally, when setting up for their

space, taking physical characteristics of the location is necessary to consider before the tents get pitched.
- Build a place for quiet and escape. Avoid constant access to media coverage of the event.
- Have a centralized place for communication and updates.

Psychological First Aid

One resource is psychological first aid (PFA; Ruzek et al. 2007), which comprises eight core actions designed at reducing posttraumatic distress and improving short- and long-term adaptive functioning for responders. The PFA Field Operations Guide is available online through the National Child Traumatic Stress Network (nctsnet.org). It is designed to be used by disaster mental health responders who may need to provide immediate support to survivors, and in any situation necessary for both individuals (Hobfoll et al. 2007; Ruzek et al. 2007) and in a small-group format.

The eight core actions, which will be discussed below, are based on the following principles (Hobfoll et al. 2007) of promoting

- A sense of safety
- Calming
- A sense of self- and community efficacy
- Connectedness
- Hope

A sense of safety can be established through the provision of information. This allows for realistic appraisal of the current situation. Ongoing information that is constructive and directive is helpful. Structuring information so people are clear as to what can happen, what has happened, and what they should do clarifies ambiguity. Another way might involve removal of individuals to a safe environment and allowing them to regain a sense of biological and psychological "normalcy." When this is possible, the risk of anxiety and stress-related symptoms can be lessened. Further, connecting individuals with their network of support in some capacity can provide relief and a sense of having some control in the situation.

The promotion of calming is aimed at the biological and emotional impact of trauma-related stress and anxiety. This can ultimately generalize to many situations, potentially causing long-term effects on functioning. Teaching that these reactions are reasonable or normal reactions to abnormal situations can help reduce anxiety and some of the catastrophic thinking that can happen when one feels overwhelmed (i.e., "I can't manage this"). Activities that teach stress management, relaxation, cognitive restructuring, and deep breathing can be easily integrated into prevention and wellness activities.

To help develop self-efficacy, it is important that individuals have some belief in their ability to manage, cope, and have solid judgment and problem solving in the situation. It is important that individuals have a sense about their capacity to manage themselves and their situation. Helping set realistic and tangible goals can help this become a data-based effort, minimizing the "I feel" aspect of evaluation.

Connectedness and social support should be promoted, and research, as discussed earlier, has repeatedly shown that there is a better outcome for those with support than for those without support. This support should be at an individual, family, and community level. Further, it provides a forum for information exchange, problem solving, norming of shared experiences, emotional understanding, and acceptance. Connectedness builds bonds that can rebuild communities and fosters a common "we" so necessary for mass recovery.

Instilling hope is the final component, and the research for purpose, meaning, and hope is vast. In this sense, the authors defined hope as a crucial component due to the increased likelihood of improved outcome because of their possessing qualities of optimism, retention of hope for the future, feeling of confidence, expectation of positive outcomes, and other hopeful beliefs and definitions (e.g., God, religion, higher power). This allows for a sense of predictability in one's self and in life in general. Encouraging positive coping and meeting challenges helps minimize avoidance, withdrawal, and isolation.

The eight core actions (Hobfoll et al. 2007; Ruzek et al. 2007) are:

1. Contact and engagement, which involve rapid establishment of contact and rapport by initiating contact that is nonintrusive, compassionate, and supportive. It is important to ask if your presence or contact is wanted.
2. Safety and comfort involve addressing immediate and ongoing safety needs, and providing for physical (including medical) and emotional comfort. Providing fact-based information is important to help mitigate the impact of stress incurred because of false or inaccurate reporting.
3. Stabilization needs as warranted. This aspect involves calming and providing containment and orienting emotionally overwhelmed survivors.
4. Information gathering related to current needs and concerns, tailored interventions, and response to those needs and concerns.
5. Practical assistance involves addressing the immediate needs and concerns of the survivors and responders.
6. Connection with social supports by providing structured opportunities for brief or ongoing contacts with sources of support.
7. Information on coping, management, and support related to stress management, stress reactions, and coping strategies that prepares them to mitigate the effects of the incident and recovery process.
8. Linkage with collaborative services allows for the survivor to have resources available for the present or future.

Critical Incident Stress Management (CISM)

Critical incident stress management (CISM) is defined by Everly and Mitchell (1997, 2000) as a system of crisis interventions that encompass the spectrum from acute crisis phase into the postcrisis phase of stress management and critical incident exposure. Although there has been debate over its efficacy, when done according to the defined structure by trained facilitators, it has been found to be very helpful, educational, preventative, and supportive. There are many articles on this topic, and the reader is directed to the International Critical Incident Stress Foundation (ICISF) for these. It has specific interventions that can be applied to individuals; to small to large groups; and to responders, families, organizations and communities. It has as its main objectives:

1. Mitigation of the impact of the event through decreasing stress reactions
2. Acceleration of recovery process, by increasing normal recovery processes, in those experiencing stress reactions
3. Restoring adaptive functioning

The International Critical Incident Stress Foundation has many resources available for reproduction and use with agencies and responders. The literature has been conflicting on the benefits of critical incident stress management (Everly and Mitchell 2000; Mitchell 2003; Ruzek et al. 2007). These criticisms were addressed very thoroughly and using a strict definition of critical incident stress management debriefing and techniques by Mitchell (2003) and Everly and Mitchell (2000). In his research analysis, Mitchell clarified that a great number of the criticisms of the model came from situations where the techniques were in fact used by untrained individuals, misapplied to people for whom the techniques were not designed, and by individuals who used them in a psychotherapy context.

From a CISM perspective, critical incidents are events that are sudden, often life-threatening and time limited, and may overwhelm the capacity to respond adaptively. Inherent in this definition, the authors point out that psychological homeostasis has been disrupted, usual coping mechanisms have failed to reestablish homeostasis, and the resulting distress has caused some impairment in functioning. In the CISM framework, crisis intervention as defined by Flannery and Everly (2000, 120) involves "provision of emergency psychological care to victims as to assist those victims in returning to an adaptive level of functioning and to prevent or mitigate the potential negative impact of psychological trauma." Mitchell (2003, p. 3) wrote, "the primary goals of the crisis intervention program entitled CISM are to mitigate the impact of a critical incident and to accelerate recovery processes of normal people who are having normal reactions to abnormal events."

Each component of the CISM crisis intervention model involves the following premise:

1. Intervene immediately to minimize the risk of maladaptive coping or responding.
2. Stabilize through mobilizing resources and supports to restore a semblance of order and routine.
3. Facilitate understanding of what has happened. Gather facts, provide information, encourage expression, helping them understand the impact of the event and aftermath.
4. Focus on problem solving as part of the effort at regaining control and self-efficacy.
5. Encourage self-reliance to restore independent functioning, practical solutions to handling the situation and establishing a normal routine and balance.

There are seven core components of CISM designed to be used in a multiple pronged component approach, where each stage has a specific desired outcome intervention, timing, activation point, and format (Everly and Mitchell 1997, 2000):

1. Precrisis preparation. Planning at this stage is primarily preventative. This stage involves stress-management training and education, anticipating crisis/disaster response. and mitigation.
2. Disaster or large-scale incidents, where it is necessary to include larger groups and types of groups (schools, churches, communities etc.). These interventions generally include demobilizations, town hall types of meetings, informational briefings, and staff advisement.
3. Defusing intervention, which is a three-phased structured small group activity. The groups should include groups of people who have the same exposure to the same event, (i.e., all first responders on the scene). This is generally provided within hours of the event. It can happen on scene or once the responders have returned to their home base or quarters. It has the purpose of assessing distress, triaging response and potential for risk from stress and exposure, and to help responders manage their reactions and responses to the situation, and to mitigate further development of problematic stress symptoms.
4. Critical incident stress debriefing (CISD), which is a seven-stage structured group discussion with individuals who are homogeneous who had exposure to the same incident, designed at mitigating acute symptoms, providing education, support, clarification of thoughts, reactions and physical responses that can occur as part of critical incident stress. This usually takes place one to ten days postincident.
5. One-to-one crisis intervention or psychological support can be offered at any point during the full range of the critical incident spectrum.
6. Family crisis intervention, education, and organizational consultation and support.
7. Follow-up and referral mechanisms for assessment and treatment if necessary for identified personnel or organizations.

Briefing and Debriefing

- Be prepared in advance about critical incident stress management and education about the response. Many agencies are writing critical incident stress management services into their demobilization policies.
- Be prepared to provide education, such as in the form of critical incident briefings, defusing (CISM), and educational briefings as part of demobilization. Demobilization is a process where the responders are provided with information about stress symptoms and potential problems that can arise from the deployment or service. It is also a brief and general education service. The International Critical Incident Stress Foundation (www.icisf.org) has some very good handouts that can be used for this purpose.
- *Equally important, before debriefing, get your people back to their people.* Let them touch their people, rest in their home, be a person, before you bring them back as a responder or to participate in debriefing. To the best extent possible, make certain they know when and where they are going back, and where their people are. When this is done, debrief them with the team they went out with, so multijurisdictional planning may be necessary to coordinate this.
- In the definition of CISM, the services are directed at groups that have had the similar experiences and exposure and are peer led.
- Before a disaster strikes, learn about any CISM teams in your region that may be available to provide services. There are federal, state, and regional teams that provide these services. In the true CISM model, the teams are made of volunteers who are also responders, so they are able to talk from a peer perspective. Some employee assistance program (EAP) services are also available, although their teams may not have peers, nor work from a traditional CISM model. The ICISF Web site provides a list of registered volunteer teams and their contact person.

Suggestions and Considerations

- Have a policy about how long a shift can or will be, and whenever possible get that implemented and routine.
- Make defusing a part of shift change whenever possible or necessary.
- Other agencies are also recognizing the importance of having disaster mental health specialists available to use during the deployment as a specific support staff whose job is to oversee the well-being of responders. Their job is to interface with each person per shift (informally), support, oversee the nourishment, convey information in and to the field if needed, and so forth. It is important to let your team know that making sure they get support at this juncture is to keep them able to do what they came to do. In some regard,

these mental health support responders become like the Father Mulcahy character on *MASH*.

- – They can be identified as support services, because these services are not designed to provide therapy, but support and psychological first aid (Ruzek et al. 2007).
- Use the buddy system for support and safety.
- Rethink the length of and frequency of shifts/in-service.
- Set up debriefings within two to seven days after leaving the scene, and debrief with the team they served with in their deployment.
- Allow people to go home and have a down-day with their people before coming back for a debriefing.
- Have a policy about how long a deployment can or will be, and whenever possible get that implemented and have a strategy for deployment into the situation and out of the situation.

There should be coordinated efforts for the critical incident stress management command staff that include briefing and information exchange per shift as well. Important information for them to discuss includes information about the changing situation, status reports, supplies, identification of potential problems and other important personnel-related data.

Other supportive staff and personnel can be made available to help with provision of supportive services to the responders, including first aid, massage, food and beverages, clean and dry clothing, and rest areas.

Conclusion

Stress is a part of our daily lives. We can manage it while we live with it. When disaster strikes it impacts the core of our existence and perception of safety and security. Different situations impact people's lives, and physical and emotional health differently. It can have a long-lasting impact on communities and the people within them. This not only affects people; it impacts economics and resources within the community.

With consideration of the aforementioned factors in mind, it is important to develop plans that anticipate and plan for a wide range of possible community disasters. Keeping this in mind, you can anticipate the needs of both the community and your responders. Although critical incidents and disasters are unpredictable, they can be planned for with some degree of anticipation of the factors discussed earlier. While we plan based on the worst we've known, we need to plan for the worst we can imagine with impact greater than we are afraid to contemplate. We also need to think outside the box, reaching for resources we've not considered, and plan to keep our personnel resources as healthy and supported as possible, not only because we need their professional skills to help us through the disaster, but also because we care about them as valuable community members.

References

American Psychological Association (APA). 2004. Mind/body health: did you know? *APA Help Center from the American Psychological Association*; http://www.apahelpcenter.org/articles/pdf.php?id=103.

Dale, K. and Fox, M. 2008. Leadership style and organizational commitment: Mediating effect of role stress. *Journal of Managerial Issues* 20(1):109–134.

DeWolfe, D. J. 2005. *Field Manual for Mental Health and Human Services Workers in Major Disasters.* DHHS Publication Number ADM 90-537. http://mentalhealth.samhsa.gov/publications/allpubs/Adm90-537/default.asp.

Everly, G. S., and Mitchell, J. T. 1997. *Critical Incident Stress Management (CISM): A New Era and Standard of Care in Crisis Intervention.* Ellicott City, MD: Chevron Publishing.

Everly, G. S., and Mitchell, J. T. 2000. The debriefing "controversy" and crisis intervention: A review of the lexical and substantive issues. *International Journal of Emergency Mental Health*, 2(4):211–225.

Fairbank, J. A., and Gerrity, E. T. 2007. Making trauma intervention principles public policy. *Psychiatry* 70(4):316–320.

Galea, S. 2007. The long-term health consequences of disasters and mass traumas. *Canadian Medical Association*, 9:176–178.

Goetzel, R. Z., Anderson, D. R., Whitmer, R. W., Ozminkowski, R. J., Dunn, R. L., and Wasserman, J. 1998. The relationship between modifiable health risks and health care expenditures: an analysis of the multi-employer HERO health risk and cost database. *Journal of Occupational and Environmental Medicine*, 40(10).

Hobfoll, S. E., Watson, P., Bell, C. C., Bryant, R. A., Brymer, M. J., Friedman, M. J., Friedman, M., et al. (2007). Five essential elements of immediate and mid-term mass trauma intervention: Empirical evidence. *Psychiatry*, 70(4):283–316.

Jackson, B. A., Baker, J. C., Ridgely, M. S., Bartis, J. T., and Linn, H. I. 2004. *Protecting emergency responders: Safety management in disaster and terrorism response*, Volume 3. DHHS (NIOSH) Publication Number 2004-114, RAND Publication Number MG-170. Cincinnati, OH: NIOSH Publications.

Mitchell, J. T. (2003). *Crisis intervention & CISM: A research summary.* International Critical Incident Stress Foundation, www.icisf.org.

Perrin, M. A., DiGrande, L., Wheeler, K., Thorpe, L., Farfel, M., and Brackbill, R. 2007. Differences in PTSD prevalence and associated risk factors among World Trade Center disaster rescue and recovery workers. *The American Journal of Psychiatry* 169(4):1385–1395.

Ray, T. K., and Sauter, S. L. 2008. *Work stress: Societal costs and organizational components.* Poster#003. The National Occupational Research Agenda (NORA), NORA Symposium 2008: Public market for ideas and partnerships, http://www.cdc.gov/niosh/nora/symp08/posters/003.html.

Rogers, J. R. 2007. Disaster response and the mental health counselor. *Journal of Mental Health Counseling* 29(1):1–3.

Ruzek, J. I., Brymer, M. J., Jacobs, A. K., Layne, C. M., Vernberg, E. M., and Watson, P. J. 2007. Psychological first aid. *Journal of Mental Health Counseling* 29(1):17–50.

Sheenhan, D. C., Everly, G. S., and Langlieb, A. 2004. Current best practices: Coping with major critical incidents. *FBI Law Enforcement Bulletin* 73(9):1–13.

Chapter 7

Developing an Emergency Operations Plan (EOP)

Michael Fagel

Contents

A jurisdiction's emergency operations plan (EOP) is a public document that does the following:

- Assigns responsibility to organizations and individuals for carrying out specific actions, at projected times and places, in emergencies that exceed the capability or routine responsibility of any one agency
- Sets forth lines of authority and organizational relationships and shows how all actions will be coordinated
- Describes how people and property will be protected in emergencies and disasters
- Identifies personnel, equipment, facilities, supplies, and other resources available within the jurisdiction, or by agreement with other jurisdictions, for use during response and recovery operations
- Identifies steps to address mitigation concerns during response and recovery activities

Why a Jurisdiction Needs an EOP

Planning to respond to emergencies and disasters is typically the responsibility of state and local governments. The elected leadership in each jurisdiction is legally responsible for ensuring that the necessary and appropriate actions are taken to protect people and property from the consequences of emergencies and disasters.

When a disaster threatens or strikes a jurisdiction, citizens expect their elected leaders to take immediate action to deal with the situation. The government is expected to marshal its resources, channel the efforts of voluntary agencies and private enterprises in the community, and solicit assistance from outside of the jurisdiction, if necessary. The development of a comprehensive, all-hazard EOP will help ensure that all government response activities are undertaken efficiently and effectively.

The Emergency Planning Process

In today's system of emergency management, local government must act to attend to the public's emergency needs. Depending on the size and nature of the emergency, state and federal assistance may be provided to the jurisdiction; however, local governments should not assume that this type of assistance will be available. Therefore, the local EOP should focus on the functions that are essential for protecting the public before and after a disaster. Minimally, these functions include providing warning, emergency public information (EPI), evacuation, and shelter. Emergency planning is not a one-time event. It is a continual cycle consisting of planning, training, exercising, and revision that takes place throughout the four phases of the emergency management cycle: mitigation, preparedness, response, and recovery.

The planning process does have a single purpose: the development and maintenance strategy for addressing critical needs in an emergency—to protect life and property.

Although the emergency planning process is cyclical, it does have a definite starting point, seen in Figure 7.1. Emergency planning begins by analyzing the hazards facing the jurisdiction. Hazard analysis is the process by which hazards that threaten the community are identified, researched, and ranked according to

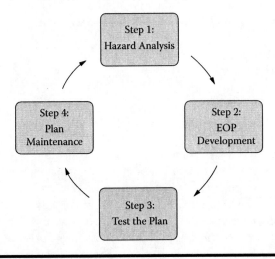

Figure 7.1   The emergency planning process.

the risks they pose, and areas and infrastructure that are vulnerable to damage from an event involving the hazards. The outcome of this step is a written hazard analysis that quantifies the overall risk to the community from each hazard.

The next step in the emergency planning process is EOP development. The outcome of this step is a completed plan, which is ready to be trained, exercised, and revised based on lessons learned from the exercises. The third step in the planning process is testing the plan through training and exercising. Exercises of varying types and complexity allow for evaluation to see what in the plan is unclear and what does not work. The outcome of this step is a series of lessons learned about weaknesses in the plan. These weaknesses can then be addressed in the final step: plan maintenance and revision. Plan maintenance and revision can be completed based on needs and resources, which may have changed since the development of the original EOP. After the EOP is developed, steps 3 and 4 repeat in a continual cycle to keep the plan up to date. If the community becomes subject to a new hazard, however, the planning team will need to revisit steps 1 and 2.

Performing a Hazard Analysis

A hazard analysis determines:

- What can occur in the community
- How often it is likely to occur
- The damage it is likely to cause
- How it is likely to affect the community
- How vulnerable the community is to the hazard

There are five steps in the hazard analysis process:

1. Identify hazards
2. Profile each hazard
3. Develop a community profile
4. Determine vulnerability
5. Create and apply scenarios

There are many potential sources of hazard information. A starting point may be local newspapers, which can provide a more comprehensive picture of the types of hazards that a community has historically faced. However, it may be necessary to check sources such as the State Department of Agriculture, Bureau of Labor Statistics, or similar agencies; the National Weather Service (NWS); local historical societies; or long-time residents. If your community has an existing hazard analysis, start by reviewing it with an eye toward what has changed in the jurisdiction since the hazard analysis was completed.

During the hazard analysis process, it is important to keep in mind that the hazards a community faces may change over time because of new mitigation measures, the opening or closing of facilities, local development activities, or terrorist threats that were not considered before the attacks of September 11, 2001.

There may be other long-term changes to investigate, as well. Changes in average temperature or rainfall and snowfall amounts may be harder to track but will certainly play an important role on the way to having a complete hazard analysis.

Developing a Hazard Profile

A hazard profile should consider four factors:

■ Magnitude
■ Frequency
■ Duration
■ Speed of onset

A hazard profile should address each hazard's magnitude—or size. How strong a hazard is and the areas that it could affect could dramatically change response plans. For example, a storm that drops two inches of rain very quickly over a small area requires a much different response than a nor'easter that drops twenty inches of rain over a four-state area.

It is also important to consider a hazard's frequency, including whether a seasonal pattern exists. In some parts of the country, thunderstorms are a near daily occurrence. On the other hand, hurricanes are a seasonal occurrence, which may or may not present a high risk to your area. Consider each hazard's duration or how long the hazard is expected to last. For example, the duration of even the most severe thunderstorm is much shorter than that of a hurricane.

Finally, consider the speed of onset of the hazard. This is important for determining the available time for issuing a warning and is also critical to the response. The amount of damage and loss of life that an extreme hazard could cause can be mitigated if emergency personnel and the public have time to take protective action.

A profile should be completed for each hazard to which the community is vulnerable, but it is important to keep in mind that some hazards pose such a limited threat that additional analysis may not be necessary. You should not, however, ignore low-risk hazards that have a high potential for damage should they occur. These low-risk hazards may not be a planning priority but should be planned for nonetheless (see Figure 7.2).

Creating a Jurisdiction Profile

After completing the hazard analysis process, it is necessary to combine hazard-specific information with a profile of the community to determine the community's

Hazard:	
Potential magnitude (Percentage of the community that can be affected): **Catastrophic:** More than 50% **Critical:** 25% to 50% **Limited:** 10% to 25% **Negligible:** Less than 10%	
Frequency of Occurrence: **Highly likely:** Near 100% probability in next year. **Likely:** Between 10% and 100% probability in next year, or at least one chance in next 10 years. **Possible:** Between 1% and 10% probability in next year, or at least one chance in next 100 years. **Unlikely:** Less than 1% probability in next 100 years.	**Seasonal Pattern:**
Areas Likely to be Affected Most:	
Probable Duration:	
Potential Speed of Onset (Probable amount of warning time): √ Minimal (or no) warning. √ 12 to 24 hours warning. √ 6 to 12 hours warning. √ More than 24 hours warning.	
Existing Warning Systems:	
Does a Vulnerability Analysis Exist?* Yes ☐ No ☐	

Figure 7.2   Hazard profile worksheet.

vulnerability to or risk of damage from the hazard. Because different communities have different profiles, vulnerabilities to the same hazard will vary. Table 7.1 summarizes key factors that are included in the community profile.

After gathering this information about the community, develop the community's jurisdiction profile by plotting vulnerable areas on a jurisdiction map. Table 7.2 shows the use of community factors in the jurisdiction profile.

Completing the Risk Analysis

After compiling the jurisdiction profile, the next step is to quantify the community's risk by merging the information. Risk is the predicted impact that a hazard would have on the people, services, and specific facilities in the community. Quantifying risk enables jurisdictions to focus on those hazards that pose the highest threat to life, property, and the environment. Quantifying risk involves:

- Identifying the elements of the community (populations, facilities, and equipment) that are potentially at risk from a specific hazard
- Developing response priorities (risk to life is always the highest priority)
- Assigning severity ratings based on the potential impact to life, essential facilities, and critical infrastructure
- Compiling risk data into the community risk profiles that show the areas of the community that are at highest risk from the hazard

In analyzing risk, it is helpful to develop response priorities, using the hierarchy in Figure 7.3 to set priorities.

Next, assign each hazard a severity rating, or risk index, that will predict, to the highest degree possible, the damage that can be expected in the community as a result of that hazard. This rating quantifies the expected impact of a specific hazard on people, essential facilities, property, and response assets. Table 7.3 is an example of severity ratings that may be used.

Develop a risk index for each hazard and assign a value to each characteristic. Use the following values:

1 = Catastrophic
2 = Critical
3 = Limited
4 = Negligible

The ratings should be assigned for each of the following types of hazard data:

- Magnitude
- Frequency of occurrence
- Speed of onset

Table 7.1 Key Factors in Creating a Community Profile

Geography	Property	Infrastructure	Demographics	Response Organizations
Major geographic features	Numbers	Utilities, construction, layout, access	Population size, distribution, concentrations	Locations
Typical weather patterns	Types	Communication system layout, features, backup	Numbers of people in vulnerable zones	Points of contact
	Ages	Road systems	Special populations	Facilities
	Building codes	Air and water support	Animal populations	Services
	Critical facilities			Resources
	Potential secondary hazards			

Table 7.2 Use of Community Factors in the Jurisdiction Profile

Type of Information	Used In
Geographic	Predicting risk factors and the impact of potential hazards and secondary hazards
Property	Projecting consequences of potential hazards to the local area Identifying available resources
Infrastructure	Identifying points of vulnerability Preparing evacuation routes, emergency communication, and projecting response and recovery requirements
Demographic	Projecting consequences of disasters on the population Disseminating warnings and public information Planning evacuation and mass care
Response organizations	Identifying response capabilities

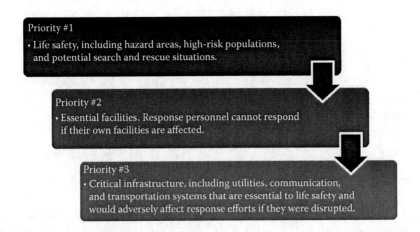

Figure 7.3 Response priorities.

Table 7.3 Hazard Severity Ratings

Severity	Expected Impact
Catastrophic	Multiple deaths Complete shutdown of critical facilities for thirty days or more More than 50 percent of property is severely damaged
Critical	Injuries and/or illnesses result in permanent disability Complete shutdown of critical facilities for at least two weeks More than 25 percent of property is severely damaged
Limited	Injuries and/or illnesses do not result in permanent disability Complete shutdown of critical facilities for more than one week More than 10 percent of property is severely damaged
Negligible	Injuries and/or illnesses treatable with first aid Minor quality of life lost Shutdown of critical facilities for two hours or less Less than 10 percent of property is severely damaged

- Community impact
- Special characteristics

Average the value of all factors to determine the overall risk level for each hazard.

The result of this process will be a prioritized list of hazards that pose the greatest threat to the community. The planning team in your community should plan for each hazard for which the risk index exceeds a predetermined threshold.

Creating Scenarios

The final step in the hazard analysis process is to create and apply scenarios for the highest-risk hazards. Scenarios should be realistic and based on the community's hazard and risk data.

To create a scenario, brainstorm to track the development of a specific type of emergency. Each scenario should describe the following:

- The initial notification that the event is occurring or is about to occur
- The potential overall impact on the community
- The potential overall impact of the event on specific community sectors
- The potential consequences, such as casualties, damages, and loss of services
- The actions and resources that would be needed to deal with the situation

Creating scenarios helps to identify situations that may exist in a disaster. These situations should be used to help ensure that a community is prepared if the hazard event occurs. Use Table 7.4 as a guide to make notes about key factors in your community.

Components of an Emergency Operations Plan

An EOP describes actions to be taken in response to natural, manmade, or technological hazards. It also details tasks to be performed by specific organizational elements at projected times and places based on established objectives, assumptions, and assessment of capabilities.

An EOP should be comprehensive. In other words, it should cover all aspects of emergency preparedness and response, and address mitigation concerns as well. It also should address all hazards and therefore be flexible enough to use in all emergencies—even unforeseen events. Finally, the EOP should be risk based. It should include hazard-specific information based on the risks that were established in the hazard analysis.

The EOP is written to provide an overview of the jurisdiction's response organization and policies. It also should provide a general understanding of the jurisdiction's approach to emergency response for all involved agencies and organizations.

An EOP consists of three parts: the basic plan, functional annexes that address the performance of a particular broad task, and hazard-specific appendices that provide additional response. So, although the basic plan provides the general approach to emergency response, it does not stand alone. Rather, it forms the basis for the remainder of the plan.

In addition, each part of the plan may have addenda in the form of standard operating procedures (SOPs), maps, charts, checklists, tables, forms, and so forth. These addenda may be included as attachments or incorporated by reference.

The Basic Plan

Although there is no standard format, for the sake of compatibility with other jurisdictions and levels of government, it is recommended that the basic plan include the following components:

- Introduction
- Purpose statement
- Situation and assumptions
- Concept of operations
- Organization and assignment of responsibility
- Administration and logistics
- Plan development and maintenance
- Authorities and references

Table 7.4   Jurisdiction Hazards

Geography	Property	Infrastructure	Demographics	Response Organizations
Major geographic features	Numbers	Utilities, construction, layout, access	Population size, distribution, concentrations	Locations
Typical weather patterns	Types	Communication system layout, features, backup	Numbers of people in vulnerable zones	Points of contact
	Ages	Road systems	Special populations	Facilities
	Building codes	Air and water support	Animal populations	Services
	Critical facilities			Resources
	Potential secondary hazards			

The Introduction

The introduction consists of five elements:

1. The *promulgation document*, which is signed by the jurisdiction's chief elected official to affirm his or her support for the emergency management agency and the planning process. It gives organizations the authority and responsibility to perform their tasks. It also mentions the tasked organizations' responsibility to prepare and maintain implementing instructions, gives notice of necessary EOP revisions, and commits to the training necessary to support the EOP.
2. The *signature page*, which is signed by all partner organizations to demonstrate their commitment to EOP implementation.
3. Dated *title page and record of changes*, which includes the date, description, and parts affected by changes to the EOP.
4. *Record of distribution*, which lists EOP recipients, and facilitates and provides evidence of EOP distribution.
5. *Table of contents*

The Purpose Statement

The purpose statement should include a broad statement about what the EOP is meant to do. It should also include a synopsis of the EOP, annexes, and appendices. The purpose statement need not be complex, but should include enough information to establish the direction for the remainder of the plan.

The Situations and Assumptions Section

The third component of the basic plan is the situation and assumptions section. The situation characterizes the planning environment, making it clear to the community why emergency planning is necessary. It draws from the hazard analysis to narrow the scope of the EOP and includes the following:

- Hazards addressed by the plan
- Relative probability and impact
- Areas likely to be affected
- Vulnerable critical facilities
- Population distribution
- Special populations
- Interjurisdictional relationships
- Maps

The assumptions statement delineates what was assumed to be true when the EOP was developed. Additionally, the assumptions statement shows the limits of the EOP, limiting liability for the jurisdiction. It may be helpful for a jurisdiction to list even obvious assumptions, such as the following:

- Identified hazards will occur
- Individuals and organizations are familiar with the EOP
- Individuals and organizations will execute their assigned responsibilities
- Assistance may be needed and, if so, will be available
- The implementation of the EOP will save lives and reduce damage

The Concept of Operations Section

The fourth component of the basic plan is the concept of operations, which explains the community's overall approach to emergency response (i.e., what, when, by whom). The concept of operations includes the following:

- The division of local, state, and federal responsibilities
- When the EOP will be activated and when it will be deactivated and—more important—by whom
- Alert levels and the basic actions that accompany each level
- The general sequence of actions before, during, and after the event
- Forms necessary to request assistance of various types

The Organization and Assignment of Responsibilities Section

The organization and assignment of responsibilities section lists, by organizations and positions, the general areas of responsibility assigned. It also identifies shared responsibilities and specifies which organization has the primary responsibility for a given function and which have supportive roles. In other words, the organization and assignment of responsibilities section specifies reporting relationships and lines of authority for an emergency response.

Administration and Logistics Section

The sixth component of the basic plan is the administration and logistics section. This section provides resource management policies and policies for augmenting response staff with public employees and volunteers, as well as a statement that addresses liability issues. It also includes the assumed resource needs for high-risk hazards, resources that are available within the community, and resources that may be available through mutual aid agreements. It is important to note, however, that the community should not rely on mutual aid agreements because neighboring jurisdictions may be faced with the same emergency or disaster that your community is facing.

Plan Development and Maintenance Section

The next component is plan development and maintenance. The responsibility for the coordination of the development and revision of the basic plan, annexes, appendices, and implementing instructions must be assigned to the appropriate persons. Therefore, this section:

- Describes the planning process
- Identifies the planning participants
- Assigns planning responsibilities
- Describes the revision cycle (i.e., training, exercising, review of lessons learned, etc.)

Authorities and References Section

The final section of the basic plan is the authorities and references section. This section cites the legal basis for emergency operations and activities, including the following:

- Laws, statutes, and ordinances
- Executive orders
- Regulations
- Formal agreements
- Predelegation of emergency authorities
- Pertinent reference materials, including related plans for other levels of government

Annexes

Annexes delineate how the community will carry out broad functions, such as issuing warnings or managing resources, in any emergency. It is important to determine early in the planning process the functions that will be included in the basic plan as annexes. When making this decision, it is important to consider the organization of the state government and that of your jurisdiction, the capabilities of your jurisdiction's emergency services agencies, and the established concept of operations. During this process, it is important to keep in mind the hazard analysis information developed for your community. What the community's planning team knows about the community's vulnerability is paramount to developing meaningful functional annexes.

Because communities vary so widely, there is no single list of functional annexes that is right for everyone. There are, however, eight core functions that FEMA recommends be addressed as annexes in every EOP:

1. The direction and control annex allows a jurisdiction to analyze the situation and decide on the best response, direct the response teams, coordinate efforts with other jurisdictions, and make the best use of available resources.
2. The communications annex provides a detailed focus on the total communications system and how it will be used.
3. The warning annex describes the warning systems in place and the responsibilities and procedures for using them.
4. The emergency public information (EPI) annex provides the procedures for giving the public accurate, timely, and useful information and instructions throughout the emergency period. It is important to note the difference between the warning annex and the EPI. Whereas the warning annex focuses on the procedures that the government uses to alert those at risk, the EPI annex deals with developing messages and accurate information, disseminating the information, and monitoring how the information is received. Because the warning system is one means for an EPI organization to get information out, the EPI annex must address coordination with those responsible for the warning system.
5. The evacuation annex describes the provisions that have been made to ensure the safe and orderly evacuation of people threatened by hazards that the jurisdiction faces.
6. The mass care annex deals with the actions that are taken to protect evacuees and other disaster victims from the effects of the disaster, including providing temporary shelter, food, medical care, clothing, and other essential needs.
7. The health and medical annex describes policies and procedures for mobilizing and managing health and medical services under emergency or disaster conditions.
8. The resource management annex describes the means, organization, and process by which a jurisdiction will find, obtain, and allocate resources to satisfy needs that are generated by an emergency or disaster.

In addition to these annexes, the planning team may want to consider annexes that make sense for the community. For example, if the community has a nuclear power plant, the planning team may want to add an annex on radiological protection. Additional functional annexes may be based on state law or jurisdictional requirements. Examples of annexes that may be added include the following:

- Damage assessment
- Search and rescue
- Emergency services
- Aviation operations

Hazard-Specific Appendices

An appendix is a supplement to an annex that adds information about how to carry out the function in the face of a specific hazard. Therefore, every annex may have several appendices, each addressing a particular hazard. The hazard-specific appendices that the planning team deems appropriate depend on the community's hazard analysis.

The decision about whether to develop an appendix rests solely with the planning team. Unlike annexes, hazard-specific appendices are not attached to the basic plan but are linked to each functional annex. Topics addressed in hazard-specific appendices include:

- Special planning requirements
- Priorities identified through the hazard analysis
- Unique characteristics of the hazard that require special attention
- Special regulatory considerations

Table 7.5 suggests appendix topics for each functional annex.

Implementing Instructions

Like the basic plan, each annex or appendix may use implementing instructions in the form of:

- Standard operating procedures (SOPs)
- Job aids
- Checklists
- Information cards
- Recordkeeping and combination forms
- Maps
- Charts
- Tables
- Forms

Implementing instructions may be included as attachments or by reference, and the planning team may use them as needed to clarify the contents of the plan, annex, or appendix. For example, the evacuation annex may be made clearer by attaching maps with evacuation routes marked. Because these routes may change depending on the location of the hazard, maps may also be included in the hazard-specific appendices to the evacuation annex. Similarly, the locations of shelters may be marked on maps supporting the mass care annex.

Standard Operating Procedures

Standard operating procedures (SOPs) are a common method of implementing instructions. SOPs provide response protocols for carrying out specific responsibilities. They describe who, what, when, where, and how. SOPs are appropriate

Table 7.5 Appendix Topics for Each Functional Annex

Annex	Appendix Topics
Direction and control	Response actions keyed to specific time periods or phases Urban search and rescue (US&R) inspection Protective gear for responders Detection equipment and techniques Laboratory analysis services Containment and cleanup teams
Communications	Provisions made to ensure that the effects of a specific hazard do not prevent or impede the ability of response personnel to communicate with each other during response operations
Warning	Hazard-specific public warning protocols Required or recommended notifications of state and federal officials
Emergency public information	Information the public will need to know about the specific hazard (e.g., special evacuation routes and shelters, in-place protective actions, etc.) The means by which information to the public will be conveyed
Evacuation	Evacuation options and timing Evacuation routes Transportation resources to support mass evacuation
Mass care	Shelter locations out of the hazard's vulnerable areas Protection of shelter occupants Food and water stocks to support extended shelter stays Capability to decontaminate people exposed to hazardous materials
Health and medical services	Unique health consequences and treatment options for people exposed to the hazard Environmental monitoring and/or decontamination requirements
Resource management	Provisions for purchasing, stockpiling, or otherwise obtaining special protective gear, supplies, and equipment needed by response personnel and disaster victims

for complex tasks requiring step-by-step instructions, tasks for which standards must be specified, and tasks for which documentation of performance protocols is required as a protection against liability.

When developing SOPs:

1. Develop a task list
2. Determine who, what, when, where, and how (Remember that "who" includes who performs the activity, to whom he or she reports, and with whom he or she coordinates)
3. Identify the steps for each task
4. Identify the standards for task completion
5. Test the procedures

Job Aids

It is important to keep the SOPs up to date through review and revision. A job aid is a written procedure that is intended to be used on the job while the task is being done. SOPs may be presented as job aids.

Job aids are also appropriate for any of the following:

- Complex tasks
- Critical tasks that could result in serious consequences
- Tasks that are done infrequently
- Procedures or personnel that change often

Job aids are also useful when conformity is needed among workers or across locations. Job aids should specify information such as the following:

- The task title
- The purpose of the task
- When to do the task
- Materials needed to perform the task
- How to perform each step of the task
- The desired results
- Standards to which the task must be performed
- How to ensure that the work is done properly

A job aid may include any of the following:

- Graphics
- Flow charts
- If–then decision tables
- Do's and don'ts

Because job aids are used during the completion of a task, they must be clear to be effective. They should use action verbs and everyday language, highlight important information, and place warnings before steps to which they apply.

Formatting is also important when creating job aids. Numbering steps and using empty space, boxes, or lines to separate steps allows users to find their place easily after looking away. Job aids may not be useful for all tasks, especially simple tasks that are performed regularly or must be accomplished quickly from memory. If a task cannot be completed while referring to a job aid at the same time, a job aid is not appropriate.

Checklists

Checklists are also useful implementing instructions. They provide a list of tasks, steps, features, contents, or other items to be checked off as completed. They often take the form of boxes to be checked off but can be developed in any form, including as rating scales. Checklists are particularly useful for tasks that are made up of multiple steps that must be completed in sequence or for when it might be necessary to document the completion of steps. Checklists may be less useful when observations must be recorded, when calculations or evaluations must be made, or when detailed instructions are required to complete the task.

Information Cards

Information cards provide information that is needed on the job in a convenient, often graphic, form. Examples include:

■ Reference forms
■ Diagrams, labeled illustrations, charts, or tables
■ Information summarized in matrix form

Things that might be presented in the form of information cards include:

■ Call-down rosters
■ Contact lists
■ Resource lists
■ Organizational charts
■ Task matrices
■ Equipment diagrams

Forms

Common forms used as implementing instructions include:

- Recordkeeping forms on which calculations, observations, or other information (e.g., damage assessment) can be recorded
- Combination forms that serve multiple functions, such as checklists with recordkeeping sections

Maps

Maps may be used as implementing instructions to highlight:

- Geographic boundaries and features
- Jurisdictional boundaries
- Locations of key facilities
- Transportation or evacuation routes

It is important to note that when using a map as an implementing instruction to show a particular feature, extraneous details are often eliminated.

Creating Effective Implementing Instructions

To be effective, implementing instructions must be appropriate for both the audience and the intended use. They must also be:

- Complete, in that they cover all of the components or steps
- Clear, concise, and easy to use; they should avoid jargon and ambiguity, be organized in a logical manner, and include instructions that identify the purpose and applicability of the particular implementing instruction
- Sufficiently detailed, in that they give all of the necessary information
- Up to date; the latest revision should be included
- Sufficient in scope; they must cover each function fully
- Identified in the EOP so that their existence is recorded

Implementing instructions should be incorporated by reference in the basic plan, annex, or appendix to which they refer.

Implementing instructions are used by all agency personnel who respond to disasters, whatever their function. They are developed at the agency level because agency personnel will be using them and, therefore, will know if they are effective. Implementing instructions used by the agency should support the agency's roles and responsibilities as described in the basic plan. For this reason, only some of the types of implementing instructions described will be useful to a particular agency, depending on its function in a response.

When developing any type of implementing instruction, the first step is to consider the job title or position and the tasks that go along with that position. You can then decide what type of implementing instructions would be most useful for those tasks.

Emergency planning is a continual cycle of planning, training, exercising, and revision that takes place throughout the four phases of the emergency cycle: mitigation, preparedness, response, and recovery.

The Emergency Planning Team

Emergency planning should be a team effort because disaster response nearly always requires coordination among many agencies and organizations, and at different levels of government. Planning as a team helps to ensure that everyone having the expertise to contribute in a response is a part of the process. Team planning also helps to ensure the buy-in of all key players or stakeholders.

Who Should Be on the Planning Team?

Different types of emergencies require different expertise and response capabilities. The specific individuals and organizations involved in a response will vary by the type of emergency or disaster. For example, law enforcement, fire services, and emergency medical services (EMS) will probably have a role to play in almost every emergency. On the other hand, hazardous materials (hazmat) personnel may or may not be involved in every incident but should be included in the planning process nonetheless, because they have specialized expertise that may be called upon in certain situations.

Table 7.6 presents an example of the types of agencies and personnel that may be involved in the planning process. All individuals do not need to be included in every aspect of the planning process, but they should be consulted in areas that affect them directly or for which they will be responsible.

Getting the Team Together

Getting all of the stakeholders in the planning process together and to take an active interest in the planning process will be an arduous task. To schedule meetings with so many participants may be even more difficult. It is critical, however, to have everyone's participation in the planning process, at least in the early stages and at critical points along the way, to draw from their expertise and ensure their buy-in to the plan.

The expertise and knowledge that participants bring of their organizations' resources is crucial to developing an accurate plan that considers the entire community's needs and the resources that could be made available in an emergency. It is definitely to the community's benefit to have the active participation of all key players.

Table 7.6 Agencies for Planning Teams

Individuals/Organizations	Contribution to the Planning Team
Chief elected official (CEO) or designee	Support for the emergency planning process, policy guidance and decision-making capability, and the authority to commit the community's resources
Fire chief or designee	Knowledge of the Incident Command System (ICS), knowledge of fire department procedures, on-scene safety requirements, hazardous materials response requirements, and search and rescue requirements; knowledge of the community's fire-related risks, and specialized personnel, and equipment resources
Police chief or designee	Knowledge of police department procedures, on-scene safety requirements, and local laws and ordinances; specialized response requirements, such as perimeter control and evacuation procedures; and specialized personnel and equipment resources
Public works director or designee	Knowledge of the community's road and utility, infrastructure, and specialized personnel and equipment resources
EMS director or designee	Knowledge of medical treatment requirements for a variety of situations, knowledge of medical treatment facility capabilities, and specialized personnel and equipment resources
Hazardous materials coordinator	Knowledge of hazardous materials that are produced, stored, or transported in or through the community; knowledge of EPA, OSHA, and DOT requirements for producing, storing, and transporting hazardous materials; knowledge of how to respond to hazardous materials and incidents
Mutual aid partners	Specialized personnel and equipment resources, and additional personnel and equipment resources
Department of health director or designee	Knowledge of community public health capabilities and limitations, familiarity with the key local healthcare providers, and specialized personnel resources

Continued

Table 7.6 (*Continued*)   Agencies for Planning Teams

Individuals/Organizations	Contribution to the Planning Team
Department of Transportation director or designee	Knowledge of the community's road infrastructure, knowledge of the area's transportation resources, familiarity with the key local transportation providers, and specialized personnel resources
U.S. Department of Agriculture director or designee	Knowledge of the area's agricultural sector and the associated risks
Tax assessor	Records of property in the community and their respective values
Building inspector	Knowledge of types of construction used in the community, knowledge of land use and land use restrictions, and records of planned development
School superintendent or designee	Knowledge of school facilities, and knowledge of hazards that directly affect schools
Voluntary agency directors	Knowledge of additional resources that can be brought to bear in an emergency, lists of shelters, feeding centers, and distribution centers, and volunteer resources
Air/seaport managers	Knowledge of risks associated with airport/ seaport operations, specialized personnel and equipment resources that can be used in an emergency
Representatives from local industry	Knowledge of hazardous materials that are produced, stored, or transported in or through the community; facility response plans, specialized personnel, and equipment resources that can be brought to bear in an emergency
Radio Amateur Civil Emergency Services (RACES)	List of RACES resources that can be used in an emergency
Social services agency representative	Knowledge of special-needs populations in the community

Table 7.6 (*Continued*) Agencies for Planning Teams

Individuals/Organizations	Contribution to the Planning Team
Veterinarian/animal shelter representative	Knowledge of the special response needs required for animals, including livestock
City or county attorney	Knowledge of legal and liability issues; see the section titled "Disaster Preparedness and the Law" in this chapter.

There are several steps that you can take to gain cooperation from participants and to gain their involvement in the planning process.

- Plan ahead—Give the planning team plenty of notice of where and when the planning meeting will be held. If time permits, survey the team members to find the time and place that will work for them.
- Provide information about team expectations—Explain why participating on the planning team is important to the participants' agencies and to the community itself. Show the team members how they will contribute to a more effective emergency response.
- Involve the chief elected official—Ask the chief elected official to sign the meeting notice. This will send a clear message that emergency planning is important to the community and that the participants are expected to attend.
- Allow flexibility in scheduling after the first meeting—Not all team members will need to attend all meetings. Task forces and subcommittees can complete some of the work. Where this is the case, gain concurrence on timeframes and milestones, but let the subcommittee members determine when it is most convenient to meet.

It also may be beneficial to talk to emergency managers from other communities to gather input on how to gain and maintain interest in the planning process.

Team Operation

Unlike working alone, working with personnel from other organizations to plan for emergencies requires some give and take—in other words, collaboration. Collaboration is the process in which people work together as a team on a common mission—in this case, the development of a community EOP. Successful collaboration requires:

- A commitment to participate in shared decision making
- A willingness to share information, resources, and tasks
- A professional sense of respect for individual team members

Collaboration can be made difficult by differences among agencies and organizations in:

- Terminology—Often, people from different organizations use different terms to mean the same thing—or use the same terms to mean different things. For example, the American Red Cross considers a house fire to be a disaster, while FEMA considers disasters to be incidents large enough in scope and impact to warrant a Presidential declaration under the Stafford Act.
- Experience—Peoples' experiences lead them to respond differently in different situations. Those whose past experiences have proven them correct (or incorrect) in a given situation may have very strong opinions about what should be done, when, and how.
- Mission—Each agency that participates in the planning process is operating toward achieving its specific mission, which may or may not be entirely consistent with that of the emergency management agency. "Personal" missions may interfere with collaboration.
- Culture—Culture entails everything about a person and the agency that he or she works for, including the potential meaning of nonverbal cues. Culture may be one of the most difficult factors to overcome during the collaboration process.

Collaboration under these conditions requires flexibility to agree on common terms and priorities, and humility to learn from others' ways of doing things.

Although collaboration among EOP planning team members may be difficult, it benefits the community by strengthening the overall response to disaster. Collaboration can:

- Eliminate duplication of services, resulting in a more efficient response
- Expand resource availability through sharing
- Enhance problem solving through the cross-pollination of ideas

Stages of Team Formation

Collaboration does not come automatically. Building a team that works well together takes time and effort, and typically evolves through five stages:

1. Forming—Individuals come together as a team. During this stage, the team members may be unfamiliar with one another and uncertain of their roles on the team.
2. Storming—Team members may become impatient, disillusioned, and may disagree.
3. Norming—Team members accept their roles and make progress toward the goal.

4. Performing—Team members work well together and make progress toward the goal.
5. Adjourning—Their task accomplished, team members may feel pride in their achievement and some sadness that the experience is ending. It is important to remember that the team should remain intact for the purpose of exercising and revising the plan.

Team Roles

To keep the team focused throughout the planning stages, it is important for team members to assume roles. Perhaps the most important role is that of team leader. The team leader initiates appropriate team-building activities that move the team through each stage and toward its goal.
Other team roles may include:

■ The task master, who identifies the work to be done and motivates the team
■ The innovator, who generates original ways to get the group's work done
■ The organizer, who helps the group develop plans for getting the work done
■ The evaluator, who analyzes ideas, suggestions, and plans made by the group
■ The finisher, who follows through on plans developed by the team

Together, all members contribute toward making the team productive.

Characteristics of an Effective Team

You will know that the planning team is on track when it displays the following characteristics:

■ Works toward a common goal
■ Accepts the leader who provides direction and guidance
■ Communicates openly
■ Resolves conflicts constructively
■ Displays mutual trust
■ Shows respect for each individual and his or her contributions

Disaster Preparedness and the Law

Why should your city or county attorney or any attorney be included on your emergency and disaster preparedness team? What does a "lawyer" know about emergency planning? What does an attorney bring to the table in the emergency planning process? Will the attorney add value to the team or be a detriment to the overall planning process?

An emergency and disaster preparedness team is assembled based upon individuals' knowledge, skills, and abilities, their expertise, and their experience with the primary goal of the team being to "do good deeds" for others in an emergency or disaster situation. Isn't it logical and prudent to have at least one member of the emergency and disaster preparedness team with a skill set that is particularly focused on the legal aspects of the team's actions and possible inactions that could inadvertently cause harm to others? From this perspective, there is a definite need for an attorney's unique legal perspective as part of any emergency and disaster preparedness team especially from the beginning of the planning process.

What an Attorney Can Provide as a Team Member

We live in a very litigious society. The good deeds that each and every member of the team wants to do to assist their community during an emergency or disaster situation may result in some form of legal action, claim, or litigation directly from their good deed, action, or even possibly from the lack of action, depending on the circumstances. An attorney on the emergency and disaster preparedness team can possibly identify potential pitfalls or potential legal risks wherein the emergency and disaster preparedness team can take proactive steps to protect against the risk, educate their members as to the risk, or even ensure against the risk to eliminate or minimize the risk from becoming a "disaster after the disaster."

An attorney on the emergency and disaster preparedness team can bring a different perspective due to the specialized skill set that was drilled into them during their legal studies. Every emergency and disaster preparedness team, no matter how well prepared, is second-guessed following any event. A skilled attorney can assist the team in analyzing the potential risks during the planning process to provide proactive alternatives to eliminate or minimize the risks. Just as there are two sides to every argument, there are two sides to every situation, depending on your perspective. A skilled attorney should be able to provide the opposing viewpoint for the emergency and disaster preparedness team, especially during the planning stage, to permit the team to carefully analyze the situation and determine if the risk is acceptable or offer alternatives to counteract the risk.

Teams are often assembled in accordance with their experience, expertise, and personality and each member is expert in his or her areas of expertise. How many team members are great at paperwork? Who on the team will double check the team's paperwork and check all of the boxes to ensure the accuracy and correctness of documents and documentation? An attorney can serve as a fail safe through which all documents can be evaluated to lessen potential liability as a result of paperwork errors.

Many of the grant proposals and other documents that may come before the emergency and disaster preparedness team may be written in legalese. The skilled attorney can interpret the legalese into a comprehensible verbiage for the emergency and disaster preparedness team members. This should permit the other team members to focus on their areas of expertise when developing the grant or other document rather than spending their time with a legal dictionary.

In addition, the media is omnipresent in any emergency and disaster situation. Although many emergency and disaster preparedness teams possess an information officer, the skilled attorney can assist in honing any statements to the press and keep the proverbial foot from entering the mouth of team members. As we are all aware, video footage from any emergency or disaster situation locks the scene in time and is often utilized to Monday-morning quarterback the actions or inactions of the team in public forums as well as in litigation in the future.

Long story short, there is very little downside to adding a city attorney or local attorney to the planning process, and the substantial "upside" is that an attorney will be able to proactively identify potential legal pitfalls during the planning process.

The Potentially Applicable Laws

Let's take a look at a few of the potentially applicable civil laws as well as the areas of potential liability or pitfalls in an emergency and disaster situation.

Federal laws (including but not limited to):
- Americans with Disabilities Act (ADA)
- Occupational Safety and Health Act (OSHA)
- Labor Management Relations Act
- Federal Service Labor–Management Relation Act
- Civil Rights Act of 1964 and 1991
- Age Discrimination in Employment Act
- Veterans' Reemployment Rights Act
- Rehabilitation Act of 1973
- Federal Anti-Discrimination Executive Orders
- Fair Labor Standards Act (FLSA)
- Minimum Wage Requirements under the FLSA
- Overtime Pay under the FLSA
- Employee Retirement Income Security Act (ERISA)
- Federal unemployment laws
- Federal privacy laws
- Family and Medical Leave Act (FMLA)

State laws (including but not limited to):
- State labor relations laws
- Antidiscrimination laws
- Protection for military personnel
- State wage and hour laws
- State safety and health laws.
- State workers' compensation laws
- Governmental immunity statutes

Local laws (including but not limited to):
- Building codes
- Jurisdictional requirements
- Land use laws
- Traffic laws

Although these are just a few of the myriad federal, state, and local laws and requirements possessing potential civil liability, there is a completely different set of laws on the criminal side. Generally, the criminal laws are applicable to individual team members' actions or inactions, rather than the overall organization; however, many of the aforementioned laws do possess a criminal component for egregious situations.

Negligence

One specific area in which an attorney can educate your emergency and disaster preparedness team is in the frequent civil actions sounding in the area of negligence. There are numerous causes of action in many areas sounding in negligence ranging from negligence hiring to negligence supervision to negligent injury.

Generally, to prove negligence, the injured party must provide:

1. Duty—The emergency and disaster team possessed a duty to act.
2. Breach—The emergency and disaster preparedness team breached this duty through their actions or inactions.
3. Causation—The actions or inactions of the emergency and disaster preparedness team caused the injury to the other party.
4. Damages—The actions or inactions of the emergency and disaster preparedness team caused damages to the other party.

Although this breakdown is provided in a simplistic manner, negligence actions are common in federal and state civil courts against public and private sector entities. For public sector emergency and disaster preparedness organizations, there is the additional issue of immunity that would need to be addressed. An attorney, as a valued member of the emergency and disaster preparedness team, can educate the team as to the requirement of these laws in order to avoid risks and areas of potential liability as well as being the well-prepared advocate in case a legal action should arise against the organization or team.

Emergency Management Phases

Let's take a look at the various stages of your emergency and disaster preparedness plan to ascertain some of the general areas where potential legal issues could arise, in which legal counsel could assist the emergency and disaster preparedness team:

Planning Stage

- There are jurisdictional issues between the state and local governmental entities.
  - Who is responsible for what activities?
- A mutual aid agreement is agreed upon between departments. One entity fails to perform the required responsibilities.
  - Who is responsible for any consequences that arise from this?

Response Stage

- The 9-1-1 system is not working.
  - Who is responsible?
  - Is the failure of the 9-1-1 system the cause of a grandma dying?
- A member of the emergency and disaster team renders first aid at the scene. The victim dies and the family blames the first responder.
  - Who may have liability?

Recovery Stage

- The team is attempting to repair a gas line and the line explodes.
  - What if a team member is injured or killed?
  - What if the neighboring house is severely damaged?
- The team member is utilizing a chain saw and the chain saw malfunctions causing severe injury.
  - Is the manufacturer responsible?
  - Your organization?
  - Is third-party recovery possible?

Other Areas of Potential Civil Liability Risks

Vehicles

- A member of your team is injured while driving a vehicle during an emergency situation.
  - Who pays the medical bills?
  - Disability for those involved both on your team and otherwise?
- A family brings a legal action for a team member killed in a vehicular accidents going to the disaster scene.
  - Can the family sue?
- A team member fails to stop at an intersection on the way to the scene and kills the other driver.
  - Can the family of the other driver sue?
  - Who would they name in the action?

- The brakes on the team member's official vehicle fail and the vehicle injures a number of children when it crashes into a pizza parlor.
  - Can the children sue or will their parents sue on their behalf?
  - Who would they sue?

Employment

- A qualified 42-year-old employment candidate applies for a job with your team and is not selected.
  - Was the candidate discriminated against due to her age?
  - Is the candidate protected under the Age Discrimination in Employment Act?
- You team adopts a "no beard" policy for safety reasons.
  - Did your team discriminate against your bearded team members?
  - Can the wearing of a beard be for religious reasons?
  - Does the wearing of a beard create a safety hazard?
- A team member is not promoted due to statements he or she has made in public with regard to a controversial animal rights issue.
  - Was the team member discriminated against due to his or her public statements or stance?
  - Are the team member's public statements considered free speech under the U.S. Constitution?

Discrimination

- A male team member continually asks a female member of the team out on a date, which she refuses.
  - Is the male team member harassing the female team member?
  - Is this actionable under Title VII of the Civil Rights Act?
  - Would the action be brought against the team member or the team/department?
- The team leader promotes a white male team member with five years experience over a Hispanic team member with ten years experience.
  - Did the team leader discriminate against the Hispanic team member?
  - Is this actionable under Title VII?
- A candidate for your team is in a wheelchair due to a permanent injury. The team decides that the candidate was not the most qualified for the position.
- Did the team discriminate against the candidate under the Americans with Disabilities Act (ADA)?

Training

- One of your team members has a heart attack and dies while training.

- Are death benefits paid under workers' compensation?
- Can the family bring a legal action?
■ A member of the team has a cold and refuses to participate in the training exercise. The team leader dismisses the team member from the team.
  - Can the team member bring a legal action against the team?
■ A team member drops a pipe on the foot of a fellow team member during a training exercise.
  - Can the injured team member sue the team member who dropped the pipe?

The previous examples are but a few situations where an attorney can be of assistance in providing guidance to the team. But these are only a few of the situations that could arise as a result of the myriad federal, state, and local laws. And these are on the civil side of the aisle. There is an entirely different set of federal, state, and local criminal laws that can impact the team or individual team members.

Conclusion

Emergency planning requires collaboration from a variety of individuals and organizations. The benefits of collaboration far outweigh the difficulties that will be faced during the planning process. Some of the benefits include elimination of duplication of services, expanded resource availability, and enhanced problem solving.

The planning process can be made easier by planning ahead, providing information about team expectations, and allowing flexibility in scheduling. It may be to your benefit to talk to emergency managers from other communities to gain their input on the planning process.

Successful team operation requires a commitment to participate in shared decision making, a willingness to share information, and a professional sense of respect for individual team members.

Team formation usually takes place through five stages: forming, storming, norming, performing, and adjourning. You will know that you have an effective planning team when members agree on and work toward a common goal, provide open communication, and display constructive conflict resolution, mutual trust, and respect for one another.

The end product of emergency planning is the emergency operations plan (EOP), a document describing how citizens and property will be protected in a disaster or emergency. There are four steps in the emergency planning process:

1. The hazard analysis, or the process by which hazards in the community are identified and ranked according to the risks that they pose to the community
2. EOP development, including the basic plan, functional annexes, hazard-specific appendices, and implementing instructions

3. Testing the plan through training and exercises to determine weaknesses and strengths in the plan
4. Plan maintenance and revision, based on needs and resources that may have changed since the development of the original EOP

Thoroughly completing each of these steps will help you develop an EOP that requires fewer changes and less significant changes following training and exercising (see Chapters 8 and 9 for more about training and exercising).

Last, when considering who should be on the emergency and disaster preparedness team, remember that a professional with specialized legal training can be a beneficial member. Lawyers do bring a lot of experience and expertise to the table. Every emergency and disaster preparedness team will encounter legal questions and situations where the appropriate legal advice can save time, money, and even lives.

References

Federal Emergency Management Agency. Plan components. *Emergency Management Guide for Business & Industry,* http://www.fema.gov/business/guide/section1c.shtm.

Tuckman, B. Developmental sequence in small groups. *Psychological Bulletin* 63(6):384–399. http://findarticles.com/p/articles/mi_qa3954/is_200104/ai_n8943663.

Chapter 8

Developing an Exercise Program

Michael Fagel

Contents

Introduction

Successful emergency responses over the years have demonstrated that exercising pays huge dividends when an emergency occurs. An exercise is a focused practice activity that places the participants in a simulated situation requiring them to function in the capacity that would be expected of them in a real event. The purpose of an exercise is to promote preparedness by testing policies and plans, and training personnel.

Why Exercise?

The goal of exercise design is to establish a comprehensive program based on a long-term, carefully constructed plan. In a comprehensive program, exercises build on one another to meet specific operational goals. The ultimate goal, of course, is to develop competency in all emergency functions.

There are two main benefits of an exercise program:

1. Individual training—Exercising enables people to practice their roles and gain experience in those roles.
2. System improvement—Exercising improves the organization's system for managing emergencies.

To get the full benefit from an exercise program, exercises need to be evaluated and recommendations acted upon. An exercise has no value unless it leads to improvement. Through exercises, organizations can:

- Test and evaluate policies, plans, and procedures
- Reveal planning weaknesses
- Reveal gaps in resources
- Improve organizational coordination and communications
- Clarify roles and responsibilities
- Improve individual performance
- Gain program recognition and support of officials
- Satisfy regulatory requirements

The goal of an exercise should always be to locate and eliminate problems before an emergency occurs. Corrective actions are an important part of exercise design, evaluation, and follow up.

Regulatory Requirements

Because the human and monetary cost of emergencies is so great, many governments, agencies, and corporate entities have mandated preparedness training and exercising. These are just a few examples of mandated exercises:

- State and local governments receiving federal funds may have to comply with certain exercise requirements. The Federal Emergency Management Agency's (FEMA's) requirements change periodically, but the program is normally structured around a four-year cycle.
- Nuclear power plants must exercise their plans yearly, conducting a full-scale exercise every two years. This exercise is evaluated by the Nuclear Regulatory Commission (NRC).
- Agencies or committees falling under the coverage of the Superfund Amendment and Reauthorization Act of 1986 (SARA) Title III must conduct a yearly exercise and evaluate their hazardous materials response and recovery plans.
- Airports, hospitals, and other healthcare facilities must conduct a full-scale exercise once every two years to maintain their certifications or licenses to operate.
- Many employers are required by the Occupational Safety and Health Administration (OSHA) to develop an emergency action plan. OSHA recommends that such plans be exercised at least annually.

Exercise Functions

In an exercise program, the emphasis is on functions rather than on types of emergencies because preparedness in those functions is common in all emergencies. Functions are actions or operations required in emergency response or recovery. FEMA defines thirteen functions in its Emergency Management Exercise Reporting System:

1. Alert notification
2. Warning
3. Communications
4. Coordination and control
5. Emergency public information (EPI)
6. Damage Assessment
7. Health and medical
8. Individual/family assistance
9. Public safety
10. Public works/engineering
11. Transportation
12. Resource management
13. Continuity of government

Each of these functions has a set of subfunctions related to it, and a community may choose to focus on some of these. For example, an emergency response focus may relate to efforts such as:

- The management and distribution of donations
- The logistics of providing needed resources
- The temporary conversion of a manufacturing process to provide emergency supplies
- How to coordinate with other organizations to provide mass care
- How employees of an organization respond to an internal emergency

The variations of these subfunctions are endless. However the entity is organized, the exercise program should identify the applicable functions and emphasize testing the operational procedures within those functions, regardless of the type of emergency.

You will need to complete a needs assessment for your jurisdiction to analyze where you may wish to focus exercise design efforts. In completing the needs assessment, you may wish to consult such resources as planning documents, demographic or corporate data, maps, and training records. See the Needs Assessment Worksheet in Figure 8.1.

The Comprehensive Exercise Program

In any discussion of emergency preparedness, the emphasis is on a comprehensive exercise program, made of progressively complex exercises, each one building on the previous one, until the exercises are as close to reality as possible.

Progressive Exercising

A progressive program has several important characteristics:

- The exercise program involves the efforts and participation of various entities, whether departments, organizations, or agencies. Through the involvement of multiple entities, the program allows the involved entities to test not only their implementation of emergency management procedures but their coordination with each other in the process.
- The program is carefully planned to achieve identified goals.
- The program is made up of a series of increasingly complex exercises.
- Each successive exercise builds upon the previous one until mastery is achieved.

Exercising Requires a Broad Commitment

When a jurisdiction engages in a progressive exercise program, the program needs to be comprehensive. The program must consider every type of responding agency and organization in the community.

It is important to keep in mind that communities are composed of more than police, fire, and public works. A progressive exercise program, therefore, requires a

Hazards

List the various hazards in your jurisdiction. What risks are you most likely to face? You can use the following checklist as a starting point. Note: If your jurisdiction has already conducted a hazard analysis, that is the best resource.

☐	Airplane crash	☐	Sustained power failure
☐	Dam failure	☐	Terrorism
☐	Drought	☐	Tornado
☐	Epidemic (biological attack)	☐	Train derailment
☐	Earthquake	☐	Tsunami
☐	Fire/Firestorm	☐	Volcanic eruption
☐	Flood	☐	Wildfire
☐	Hazardous material spill/release	☐	Winter storm
☐	Hostage/Shooting	☐	Workplace violence
☐	Hurricane	☐	Other _____
☐	Landslide/Mudslide	☐	Other _____
☐	Mass fatality incident	☐	Other _____
☐	Radiological release	☐	Other _____

Secondary Hazards

What secondary effects from those hazards are likely to impact your jurisdiction?

☐	Communication system breakdown
☐	Power outages
☐	Transportation blockages
☐	Business interruptions
☐	Mass evacuations/displaced population
☐	Overwhelmed medical/mortuary services
☐	Other _____
☐	Other _____
☐	Other _____

Figure 8.1   Needs assessment worksheet.   *Continued*

Hazard Priority

What are the highest priority hazards? Consider such factors as:

Frequency of occurrence.

Relative likelihood of occurrence.

Magnitude and intensity.

Location (affecting critical areas or infrastructure).

Spatial extent.

Speed of onset and availability of warning.

Potential severity of consequences to people, critical facilities, community functions, and property.

Potential cascading events (e.g., damage to chemical processing plant, dam failure).

#1 Priority hazard:

#2 Priority hazard:

#3 Priority hazard:

Area

What geographic area(s) or facility location(s) is(are) most vulnerable to the high priority hazards?

Plans and Procedures

What plans and procedures—emergency response plan, contingency plan, operational plan, standard operating procedures (SOPs)—will guide your jurisdiction's response to an emergency?

Figure 8.1 (*Continued*)　**Needs assessment worksheet.**

Functions

What emergency management functions are most in need of rehearsal? (e.g., What functions have not been exercised recently? Where have difficulties occurred in the past?) You can use the following checklist as a starting point.

☐	Alert notification (emergency response)	☐	Public safety
☐	Warning (public)	☐	Public works/engineering
☐	Communications	☐	Transportation
☐	Coordination and control	☐	Resource management
☐	Emergency Public Information (EPI)	☐	Continuity of government or operations
☐	Damage assessment	☐	Other _____
☐	Health and medical	☐	Other _____
☐	Individual/family Assistance	☐	Other _____

Participants

Who (agencies, departments, operational units, personnel) needs to participate in an exercise? For example:

• Have any entities updated their plans and procedures?

• Have any changed policies or staff?

• Who is designated for emergency management responsibility in your plans and procedures?

• With whom does your organization need to coordinate in an emergency?

• What do your regulatory requirements call for?

• What personnel can you reasonably expect to devote to developing an exercise?

Program Areas

Mark the status of your emergency program in these and other areas to identify those most in need of exercising.

	New	Updated	Exercised	Used in Emergency	N/A
Emergency Plan					
Plan Annex(es)					
Standard Operating Procedures					
Resource List					

Figure 8.1 (*Continued*)   Needs assessment worksheet.

	New	Updated	Exercised	Used in Emergency	N/A
Maps, Displays					
Reporting Requirements					
Notification Procedures					
Mutual Aid Pacts					
Policy-Making Officials					
Coordinating Personnel					
Operations Staff					
Volunteer Organizations					
EOC/Command Center					
Communication Facility					
Warning Systems					
Utility Emergency Preparedness					
Industrial Emergency Preparedness					
Damage Assessment Techniques					
Other:					

Past Exercises

If your jurisdiction has participated in exercises before, what did you learn from them, and what do the results indicate about future exercise needs? For example, consider the following questions:

- Who participated in the exercise, and who did not?
- To what extent were the exercise objectives achieved?
- What lessons were learned?
- What problems were revealed, and what is needed to resolve them?
- What improvements were made following past exercises, and have they been tested?

Figure 8.1 (*Continued*)   Needs assessment worksheet.

commitment from various agencies and organizations to participate in increasingly challenging exercises over a period of time to address the entire emergency management system rather than a single problem. The community must consider the role of each entity participating in the exercise, and must secure the commitment of all these entities to build a coordinated, effective response.

Exercising Requires Careful Planning

Exercises require careful planning around clearly identified goals. Only through identifying exercise goals, then designing, developing, and conducting the exercise and analyzing the results, can those who are responsible for emergency planning be sure of what works and what does not.

Exercises should be organized to increase in complexity. For example, start with a tabletop discussion, move to a functional exercise, and finish with a full-scale exercise. Each exercise builds on previous exercises using more sophisticated simulation techniques and requiring more preparation time, personnel, and planning.

Rushing into a full-scale exercise too quickly can open the door to potential failure because shortfalls in the planning process have not been identified through less complicated and less expensive exercises. Each type of exercise will be explained in more detail later in this chapter.

An important advantage to building incrementally to a full-scale exercise is that successful exercises breed successful exercises. In other words, the more successes in an exercise program, the more confidence in the process the participants will have, and the more motivated they will be for the next exercise. In turn, officials and stakeholders will be more willing to commit resources, and operating skills will improve. Everybody wins.

Who Participates in the Exercise Program?

For a community-wide exercise program, the jurisdiction decides what agencies, organizations, and stakeholders participate in each exercise. Participants are further determined by the nature and size of the exercise. Larger exercises would include all of the participants who would have responsibilities in a real emergency. Smaller exercises, which focus on a limited set of functions, would limit the number of participants. On one end of the spectrum, a tabletop exercise might involve only key decision makers; on the other end of the spectrum, a full-scale exercise might involve the entire community.

Some types of exercises have additional participant requirements. For example, a functional exercise involves not only players but simulators, controllers, and evaluators.

Activities Included in the Exercise Program

There are five main types of activities in a comprehensive exercise program:

1. Orientation seminar
2. Drill
3. Tabletop exercise
4. Functional exercise
5. Full-scale exercise

As discussed, these activities build from simple to complex, from narrow to broad, and from least expensive to most costly. When carefully planned to achieve specified objectives and goals, this progression of exercise activities provides an important element of an integrated emergency preparedness system.

Orientation Seminar

As the name suggests, an orientation seminar is an overview or introduction, with the purpose of familiarizing participants with roles, plans, procedures, or equipment. It can also be used to resolve questions of coordination and assignment of responsibilities. Key characteristics of an orientation seminar are summarized in Table 8.1.

There are no hard and fast rules for conducting an orientation. Its purpose will determine its format. A few helpful guidelines for conducting a seminar follow.

Orientation Seminar Guidelines

- Be creative—You can use various discussion and presentation methods. Think of interesting classes that you have attended in other subjects, and borrow the techniques of good teachers and presenters.
- Get organized and plan ahead—Even though orientation seminars are less complex than other exercises, it is no time to find yourself unprepared.
- Be ready to facilitate a successful orientation seminar—Discourage long tirades, keep exchanges crisp and to the point, focus on the subject at hand, and help everyone feel good about being there.

Drills

A drill is a coordinated, supervised exercise activity, normally used to test a single specific operation or function. With a drill, there is no attempt to coordinate organizations or fully activate the emergency operations center (EOC). Its role in an exercise program is to practice and perfect one small part of the response plan and

Table 8.1 Orientation Seminar Characteristics

Format	The orientation seminar is a very low-stress event, usually presented as an informal discussion in a group setting. There is little or no simulation. A variety of seminar formats can be used, including: • Lecture • Discussion • Slide or video presentation • Computer demonstration • Panel discussion • Guest lecturers
Applications	The orientation seminar can be used for a wide variety of purposes, including: • Discussing a topic in a group setting • Introducing something new • Explaining existing plans to new people • Introducing a cycle of exercises or preparing participants for success in more complex exercises • Motivating people for participation in subsequent exercises
Leadership	Orientations are led by a facilitator, who presents information and guides the discussion. The facilitator should have some leadership skills, but very little other training is required.
Participants	A seminar may be cross-functional, involving one or two participants for each function or service being discussed. It may also be geared toward several participants from a single organization or department.
Facilities	A conference room or any other fixed facility may be used, depending on the purposes of the orientation.
Time	Orientations should last a maximum of one to two hours.
Preparation	An orientation is quite simple to prepare and conduct. Two weeks' preparation is usually sufficient. Participants need no preparation or previous training.

help prepare for more extensive exercises. The effectiveness of a drill is its focus on a single, relatively limited portion of the overall emergency management system. Key characteristics of drills are summarized in Table 8.2.

How a drill is conducted depends on the type of drill. Drills can range from simple operational procedures to more elaborate communication and command post drills. For example, a command post drill would require the participants to report to a drill site where a visual narrative would be displayed in the form of a mock emergency. Equipment, such as command boards and other needed supplies, would be made available.

Table 8.2 Key Characteristics of Drills

Format	A drill involves actual field or facility response for an EOC operation. It should be as realistic as possible, employing any equipment needed for the function being tested.
Applications	Drills are used to test a specific operation. They are also used to provide training with new equipment, to develop new policies or procedures, or to practice and maintain current skills. Some examples of drills run by different organizations are: • EOC: Call-down procedures • Public works: Locating and placing road barriers under time constraints • Chemical plant: Evacuation and isolation of spill area and valve system shutoff. • Airport: Response to a given runway in a specified amount of time • Military: Activation and mobilization drill
Leadership	A drill can be led by a manager, supervisor, department head, or exercise designer. Staff must have a good understanding of the single function being tested.
Participants	The number of participants depends on the function being tested. Coordination, operations, and response personnel could be included.
Facilities	Drills can be conducted within a facility, in the field, or at the EOC or other operating center.
Time	One-half to two hours is usually required.
Preparation	Drills are one of the easiest kinds of exercise activities to design. Preparation may take about a month. Participants usually need a short orientation beforehand.

Given the variety of functions that may be drilled, there is no set way to run a drill. However, the following are some general guidelines.

Drill Guidelines

- Prepare—If operational procedures are to be tested, review them beforehand. It is also important to review safety precautions.
- Set the stage—It's always good to begin with a general briefing, which sets the scene and reviews the drill purpose and objectives. Some designers like to set the scene using films, slides, or videotapes.
- Monitor the action—After a drill has been started, it will usually continue under its own steam. If you find that something you wanted to happen is not happening, you might want to insert a message to trigger that action.

Tabletop Exercises

A tabletop exercise (discussed in detail in Chapter 9) is a facilitated analysis of an emergency situation in an informal, stress-free environment. It is designed to elicit constructive discussion as participants examine and resolve problems based on existing operational plans and identify where those plans need to be refined. The success of a tabletop exercise is largely determined by group participation in the identification of problem areas.

There is minimal attempt at simulation in a tabletop exercise. Equipment is not used, resources are not deployed, and time pressures are not introduced. Key characteristics of the tabletop exercise are summarized in Table 8.3.

Functional Exercises

A functional exercise (discussed in detail in Chapter 9) is a fully simulated interactive exercise that tests the capability of an organization to respond to a simulated event. The exercise tests multiple functions of the organization's operational plan. It is a coordinated response to a situation in a time-pressured, realistic simulation.

A functional exercise focuses on coordination, integration, and interaction of an organization's policies, procedures, roles, and responsibilities before, during, or after the simulated event. A discussion of how to conduct a functional exercise will appear later in this chapter. The key characteristics of a functional exercise are summarized in Table 8.4.

Full-Scale Exercise

A full-scale exercise (discussed in detail in Chapter 9) simulates a real event as closely as possible. It is an exercise designed to evaluate the operational capability of emergency management systems in a highly stressful environment that simulates

Table 8.3   Key Characteristics of a Tabletop Exercise

Format	The exercise begins with the reading of a short narrative, which sets the stage for the hypothetical emergency. Then, the facilitator may stimulate discussion in two ways: • Problem statements—Problem statements may be addressed, either to individual participants or to participating departments or agencies. Recipients of problem statements then discuss the actions that they might take in response. • Simulated messages—These messages are more specific than problem statements. Again, the recipients discuss their responses.
	In either case, the discussion generated by the problem focuses on roles, plans, coordination, the effect of decisions on other organizations, and similar concerns. Often maps, charts, and packets of materials are used to add to the realism of the exercise.
Applications	Tabletop exercises have several important applications. They: • Lend themselves to low-stress discussions of coordination and policy • Provide a good environment for problem solving • Provide an opportunity for key agencies and stakeholders to become acquainted with one another, their interrelated roles, and their respective responsibilities. • Provide good preparation for a functional exercise.
Leadership	A facilitator leads the tabletop discussion. This person decides who gets the messages or problem statements, calls on others to participate, asks questions, and guides the participants toward sound discussion.
Participants	The objectives of the exercise dictate who should participate. The exercise can involve many people and many organizations. This may include all entities that have a policy, planning, or response role.
Facilities	A tabletop exercise requires a large conference room where participants can surround a table.

Continued

Table 8.3 (*Continued*) Key Characteristics of a Tabletop Exercise

Time	A tabletop exercise usually lasts from 1 to 4 hours but can be longer. Discussion times are open-ended, and participants are encouraged to take their time in arriving at in-depth decisions. Although the facilitator maintains an awareness of time allocation for each area of discussion, the group does not have to complete every item for the exercise to be a success.
Preparation	It typically takes about a month to prepare for a tabletop exercise. Preparation also usually requires at least one orientation and sometimes one or more drills.

actual response conditions. To accomplish realism requires mobilization and actual movement of emergency personnel, equipment, and resources. Ideally, the full-scale exercise should test and evaluate most functions of the EOP.

A full-scale exercise differs from a drill in that it coordinates the actions of several entities, tests several emergency functions, and activates the EOC or other operating center. Realism is achieved through:

- On-scene action and decisions
- Simulated victims
- Search and rescue requirements
- Communication devices
- Equipment deployment
- Actual resource and personnel allocation

Key characteristics of a full-scale exercise are summarized in Table 8.5.

Each of the five activities just described plays an important part in the overall exercise program.

Building an Exercise Program

A progressive exercise program involves the combined efforts of many agencies, departments, or other entities in a series of activities that increase in complexity until mastery is achieved.

Building an exercise program is a little like building a single exercise, except that activities take place on a much larger scale. Plans are developed by a team and are based on careful examination of the EOP.

The development of an exercise program has many facets, including:

- Analysis of capabilities and costs
- Scheduling of tasks

Table 8.4   Key Characteristics of the Functional Exercise

Format	This is an interactive exercise, similar to a full-scale exercise without the equipment. It simulates an incident in the most realistic manner short of moving resources to an actual site. A functional exercise is:
	• Geared for policy, coordination, and operations personnel—the players in the exercise—who practice responding in a realistic way to carefully planned and sequenced messages given to them by simulators.
	• A stressful exercise because the players respond in real time, with on-the-spot decisions and actions. All of the participants' decisions and actions generate real responses and consequences from other players.
	• Complex. Messages must be carefully scripted to cause participants to make decisions and act on them. This complexity makes a functional exercise difficult to design.
Applications	Functional exercises make it possible to test several functions and exercise several agencies or departments without incurring the cost of a full-scale exercise. A functional exercise is always a prerequisite to a full-scale exercise.
Leadership and participants	Functional exercises are complex in their organization of leadership and the assignment of roles. The following general roles are used:
	• Controller—Manages and directs the exercise
	• Players—Participants who respond as they would in a real emergency
	• Simulators—Assume external roles and deliver planned messages to the players
	• Evaluators—Observers who assess performance
Facilities	Functional exercises are usually conducted in the EOC or other operating center. Ideally, people gather where they would actually operate in an emergency. Players and simulators are often seated in separate areas or rooms. Realism is achieved by the use of telephones, radios, televisions, and maps.
Time	A functional exercise requires from 3 to 8 hours, although it can run a full day or even longer.

Continued

Table 8.4 (*Continued*)   Key Characteristics of the Functional Exercise

Preparation	Plan 6 to 18 months or more to prepare for a functional exercise, for several reasons:
	• Staff members need considerable experience with the functions being tested.
	• The exercise should be preceded by lower-level exercises, as needed.
	• The controller, evaluators, and simulators require training.
	• The exercise may require a significant allocation of resources and a major commitment from organizational leaders.

- Public relations efforts
- Development of a long-term plan

Careful work on the long-term plan will carry over into the design of individual exercises.

The Exercise Planning Team

A comprehensive exercise plan requires the combined efforts of many people. For a community program, the team should consist of representatives from every major government agency in the jurisdiction, and from private and volunteer organizations large enough to have exercise mandates:

- Fire department
- Sheriff
- Public works
- Hospital
- Airport
- Schools
- Volunteer organizations

The emergency manager and other emergency personnel should take the lead, and representatives should then meet to analyze what they need to do to support each other. This teamwork will help establish relationships among participants' departments or organizations.

Goal Setting

Because a comprehensive exercise program usually extends over several months, it is important to set long-term goals or develop a mission statement. Without this, the program is likely to lack focus and continuity.

Table 8.5   Key Characteristics of a Full-Scale Exercise

Format	The exercise begins with a description of the event, communicated to responders in the same manner as would occur in a real event. Personnel conducting the field component must proceed to their assigned locations, where they see a "visual narrative" in the form of a mock emergency. From then on, actions taken at the scene serve as input to the simulation taking place at the EOC or operating center.
Applications	Full-scale exercises are the ultimate in testing of functions — the "trial by fire." Because they are expensive and time consuming, it is important that they be reserved for the highest-priority hazards and functions.
Leadership and participants	One or more controllers manage the exercise, and evaluators are required. All levels of personnel take part in a full-scale exercise: • Policy personnel • Coordination personnel • Operations personnel • Field personnel
Facilities	The event unfolds in a realistic setting. The EOC or other operations center is activated, and field command posts may be established.
Time	A full-scale exercise may be designed to be as short as 2 to 4 hours, or to last as long as 1 or more days.
Preparation	Preparation for a full-scale exercise requires an extensive investment of time, effort, and resources — twelve to eighteen months or longer to develop a complete exercise package. This timeframe includes multiple drills and preparatory tabletop and functional exercises. In addition, personnel and equipment from participating agencies must be committed for a prolonged period of time.

Scheduling and Sequences

After the preliminary steps (organizing the planning team and establishing a mission statement and goals) have been taken, the hard work of drawing up a plan can take place. Developing the exercise program involves:

- Laying out a series of exercises that can meet the needs of the various participating entities
- Organizing them into a workable sequence and time schedule

Plan Format

An exercise program plan can use any format, but it should include the following elements:

- A timeframe
- A problem statement
- Long-range goals
- Functional objectives
- A schedule
- Exercise descriptions, including:
  - Type of exercise
  - Participants
  - Purpose
  - Rationale

A sample plan format is shown in Table 8.6. This is a hypothetical example of one community's exercise plan.

Working from the needs assessment that you completed for your jurisdiction, develop a plan for a comprehensive exercise program to address those needs. Be sure to include key elements. See Figure 8.2. If this format doesn't work for you, change it to meet your needs.

The Exercise Process

When an exercise proceeds smoothly, it looks easy, but there is much more to it than the time spent in the exercise itself. A great deal of thought and planning preceded the exercise, and more work will follow.

An exercise is not an independent activity with a clearly marked beginning and ending. Instead, it should be seen as part of a complex process that involves a number of tasks. All of these tasks are interrelated; they affect not only the success of the current exercise, but the success of future exercises.

Table 8.6   Sample Plan Format

Plan Format	Sample Plan: Comprehensive Exercise Program
Timeframe	The exercise program extends over an eighteen-month period.
Present problems	This program has been formulated to address problems arising as a result of rapid population growth. According to experts, possibilities for a mass casualty incident are increasing. Personnel involved in the functions listed below have not been tested in the last year.
Long-range goal	To work toward a full-scale exercise testing all important functions in the context of a mass casualty incident. This will satisfy FEMA requirements and full-scale exercise requirements for the hospital and airport by involving these entities.
Functions to be tested	Health and Medical, Public Information, Coordination and Control (EOC Operations, Incident Command) 1. To determine the adequacy of plans and procedures within the following functional areas to handle a mass casualty incident: Health and Medical, Public Information, Coordination and Control (EOC Operations, Incident Command) 2. To test the ability of the above-named functional areas to communicate and coordinate their response efforts during a mass casualty incident 3. To test the ability to obtain adequate resources (locally through mutual aid agreements) in the above-named functional areas to handle a mass casualty incident
First Month	Exercise: Orientation For: Emergency management staff and heads of various agencies: Mental health association, state funeral director, county coroner, county fire, county police Purpose: To review new plans and procedures for dealing with mass casualty incidents Rationale: Inform those who are unaware of plans and gain support and additional input from department leaders

Continued

Table 8.6 (*Continued*)   Sample Plan Format

Second Month	Exercise: Orientation
	For: Emergency management staff and heads of various agencies: Fire, police staff, county public information officer (PIO)
	Purpose: To review new plans for mass casualty incidents with responders
	Rationale: Gain support and additional input from first responders and acquaint them with leadership's plans
Fourth Month	Exercise: Training course with functional exercise
	For: Responders and incident commanders; emergency management staff; various chiefs, captains, lieutenants from fire and police; emergency medical services (EMS); mental health; Radio Amateur Civil Emergency Services (RACES); funeral directors; county coroner; county PIO
	Purpose: To provide training in field mass casualty incident response
	Rationale: This is a training session in the FEMA Field Mass Casualty Incident Response Course. This course provides an excellent overview of specific needs related to a mass fatality incident. The course culminates in a functional activity.
Seventh Month	Exercise: Drill
	For: Fire, police, EMS, coroner, funeral directors
	Purpose: To set up the Incident Command System (ICS) for responding to mass fatality incidents
	Rationale: Establish ICS to support needed functions and tasks.
Eighth Month	Exercise: Drill
	For: PIO, fire, police, emergency manager
	Purpose: To set up joint information center (JIC)
	Rationale: To acquaint participants with the PIO function and JIC operations, test equipment and lines of communication
Ninth Month	Exercise: Drill
	For: Mental health, funeral directors, PIO, clergy, emergency manager
	Purpose: To set up a family assistance center
	Rationale: To acquaint participants with the office equipment and test roles as support to the victims' families

Table 8.6 (*Continued*)   Sample Plan Format

Eleventh Month	Exercise: Tabletop exercise
	For: Incident Command, PIO, police, fire, EMS
	Purpose: To pull together the three functions tested in the previous drills in the context of a mass casualty incident as the result of a hotel fire
	Rational: Address and resolve potential communication and coordination problems among the incident command, PIO, police, fire, and EMS before the functional exercise
Fourteenth Month	Exercise: Functional exercise
	For: Communications, coordination and control, ICS and EOC, PIO, health and medical
	Purpose: To test additional functions for mass fatality in the context of a plane crash; emergency public information (EPI) effectiveness, health and medical mass casualty, coordination and control, ICS, and EOC operations
	Rationale: Identify preliminary shortfalls and test overall coordination before full-scale exercise
Fifteenth Month	Exercise: Tabletop exercise
	For: Communications, coordination and control, ICS and EOC, PIO, health and medical
	Purpose: To correct and retest problems identified in preceding functional exercise
	Rationale: Work out potential problems discovered in the previous functional exercise and make adjustments necessary before the full-scale exercise
Eighteenth Month	Exercise: Full-scale exercise: Airplane Crash
	For: All agency heads and responders
	Purpose: To test all functions in the context of a mass casualty airplane crash
	Rationale: This exercise fulfills full-scale requirement for FEMA, Federal Aviation Administration (FAA) requirements for airports, and Joint Commission on Accreditation of Healthcare Organizations (JCAHO) certification for the hospital

Comprehensive Exercise Program Planning Worksheet	
Timeframe:	
Present Problems:	
Long-Range Goal(s):	
Functional Objectives:	
Month:	
Exercise:	
For:	
Purpose:	
Rationale:	
Month:	
Exercise:	
For:	
Purpose:	
Rationale:	

Figure 8.2 Developing a comprehensive exercise program plan.

In preparation for launching an exercise program and designing individual exercises, it is important to have a clear vision of the entire exercise process. What follows are three graphic representations of the process:

- ■ Organized by task sequence
- ■ Organized by task categories and phase
- ■ Organized by major accomplishments

Sequence of Main Tasks

In Figure 8.3, the main tasks are shown in their approximate sequence. This chart may help you get a good mental picture of the entire sequence. It's also a good place to start in creating a more detailed schedule of tasks.

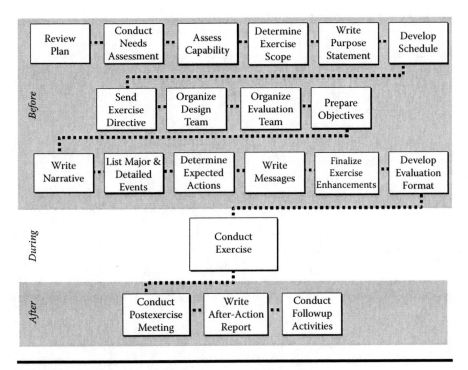

Figure 8.3 Sequence of tasks for a successful exercise.

Categories of Tasks

Another way to look at the exercise process is by organizing the tasks into two dimensions:

1. Exercise phase (preexercise, exercise, and postexercise)
2. Type of task (those related to design and those related to evaluation)

Table 8.7 illustrates this type of organization.

Major Task Accomplishments

One of the simplest ways to envision the exercise process is by major accomplishments. As shown in Figure 8.4, the process can be factored into five major accomplishments that make up the design cycle:

■ Establishing the base
■ Exercise development
■ Exercise conduct

Table 8.7 Categories of Tasks

	Preexercise Phase	Exercise Phase	Postexercise Phase
Design	• Review plan • Assess capability • Address costs and liabilities • Gain support/issue exercise directive • Organize design team • Draw up exercise schedule • Design exercise	• Prepare facility • Assemble props and other enhancements • Brief participants • Conduct exercise	
Evaluation	• Select evaluation team leader • Develop evaluation methodology • Select and organize evaluation team • Train evaluators	• Observe assigned objectives • Document actions	• Assess achievement of objectives • Participate in postexercise meetings • Prepare evaluation report • Participate in follow-up activities

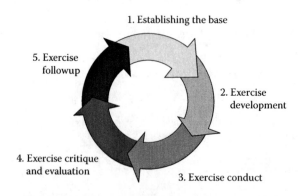

Figure 8.4 The five major task accomplishments.

- Exercise critique and evaluation
- Exercise follow-up

As shown in the graphic, each accomplishment is an outgrowth of a set of specific tasks and subtasks. The process is circular, with the results of one exercise providing input for the next.

Flexibility Is Key

The exercise process applies no matter what level of exercise is being designed or what organization is using it. Whether you are located in a large jurisdiction with extensive resources or a smaller community with meager resources, you can use this process.

The underlying premise is that this process must be flexible enough to meet the unique exercise needs of the organization that is using it. Therefore, as you consider each task, it's important to remember that each task must be designed, tailored, and applied in a manner that suits your jurisdiction's specific objectives and capabilities.

Accomplishment 1: Establishing the Base

Exercises are designed to motivate personnel to think or act as they would in a real event. Establishing the base is basically laying the groundwork for the exercise to ensure that motivation occurs. Getting ready for the exercise involves:

Reviewing the current plan—What does it tell you about ideal performance (i.e., how does your jurisdiction implement policies and procedures in the event of an emergency)?

Conducting a needs assessment—What are the risks that your jurisdiction faces, and where do you need to focus your training efforts?

Assessing the jurisdiction's capability to conduct an exercise—What resources can your jurisdiction draw upon to design and implement an exercise?

Defining the exercise scope—How should your jurisdiction limit this exercise?

Selecting the exercise type—What type of exercise best meets your jurisdiction's training needs within the available resources?

Addressing the costs and liabilities—What will the exercise cost in terms of funding, human resources, and jurisdictional liability?

Developing a statement of purpose—What do you expect your jurisdiction to gain from the exercise?

Gaining support and announcing the exercise—How can you obtain the support of those in positions of authority in your jurisdiction? How can you use that support to garner support from your participants?

Additional "groundwork" tasks include organizing a design team and developing a work plan and schedule.

Establishing the base need not take a long time and can often be done largely at your desk or with the help of a few other people. Some of these tasks will be revisited shortly.

Accomplishment 2: Exercise Development

Exercises, particularly tabletop, functional, and full-scale exercises, are developed by following an eight-step process:

1. Assess needs
2. Define scope
3. Write a statement of purpose
4. Define objectives
5. Compose a narrative
6. Write major and detailed events
7. List expected actions
8. Prepare messages

Duplicate Steps

Performing a needs assessment, defining scope, and statement of purpose are important aspects of developing the overall exercise program. They crop up again and again when establishing the base for an individual exercise. And here they appear as part of exercise development. Does this mean that we just keep repeating the same steps? Certainly not. Sometimes (as in a simple exercise), the effort that you put into these steps in the early stages will suffice—you'll already have done the work when you get to development. Other times (as with a more complex exercise), you may need to come back to the needs assessment, scope, and purpose statement, and build on them, or refine them further for the particular exercise. The more complex the exercise, the more detailed the planning and development tend to be.

Accomplishment 3: Exercise Conduct

The day of the exercise is the culmination of your planning. Here are a few suggestions for ensuring that your exercise is a success:

1. Be clear—The success of an exercise depends largely on the participants having a clear and consistent understanding of what is expected of them. Many exercises fail because ground rules or simulation techniques to be used during the exercise are inadequately explained.

2. Sustain action—Sustaining action through messages is one way to meet the key objectives of the scenario. Messages help keep all participants active throughout the entire exercise. This requires careful monitoring and control of the message flow throughout the exercise.
3. Foster realism—Participants should be encouraged to treat the simulated emergency situations as if they were actually occurring.
4. Establish timelines—The establishment of valid timelines helps keep what is occurring in valid order.
5. Review emergency call-off procedures—To ensure that all participants understand the procedure for a real emergency call-off, they should be briefed before the exercise begins.
6. Capitalize on problem situations—Situations that can cause stoppage in an exercise also have value because they place added stress on the system. Often, they will more effectively test the organization's capability to cope in times of real emergency.

Accomplishment 4: Exercise Evaluation and Critique

An exercise has no value unless it leads to improvement in the overall response system; therefore, an important part of any exercise is evaluating how well the objectives have been achieved. The objective might relate to issues, such as:

■ Needed improvements in the plan
■ Needed improvements in the emergency management system
■ Personnel training
■ Overcoming staffing deficiencies

The extent and depth of the evaluation are determined by the participating organizations. Controllers' evaluations and observations may suffice for some exercises; whereas, for other exercises, additional analysis from objective observers may be required.

Critiques and reports analyze and explain the attainment or nonattainment of the exercise objectives and provide recommendations for addressing any deficiencies. Evaluators should be thoroughly familiar with the community's EOP and the area that they are evaluating. Strategies for evaluating an exercise will be discussed in greater depth later in the chapter.

Accomplishment 5: Exercise Follow-Up

Follow-up is one of the most important but also one of the most neglected areas of the exercise process. Similarly, recommendations without follow up will keep you from getting the full benefit from an exercise. Some strategies for following up an exercise are the following:

- Assign responsibility—Clearly assign tasks and schedules, and designate responsibility for each recommended improvement.
- Monitor—Establish a monitoring plan to track the progress of implementing recommended improvements.
- Complete the cycle—Build the testing of improvement into the next exercise. This is perhaps the surest way to make certain that they are implemented.

It is probably not necessary to retest every objective fully. Rather, select a few recommendations that would illustrate improvements and include those in a future exercise.

This overview has presented the exercise process in a nutshell. As mentioned earlier, a more detailed discussion of many aspects of the exercise process will appear later in the chapter.

Reviewing the Current Plan

The EOP describes how personnel in the organization should respond in the event of an emergency. It answers the question, "What resources, personnel, and procedures will be used to resolve problems created by an emergency?"

Your plan may be called by various names (EOP, area contingency plan, etc.) but for the purpose of conformity, these documents will simply be referred to as the *emergency plan*.

Examining the emergency plan will help you identify problems, select an exercise, define its purpose, and formulate objectives. Whereas exercises test performance, it is the emergency plan that describes the ideal performance. To use the emergency plan as the base document, it is imperative that you first become familiar with it.

What to Look for in the Emergency Plan

While reviewing the plan, it is important to ask yourself the following questions:

- What responses are currently planned?
- What resources, personnel, and procedures will be used to resolve problems?
- Are they different for various agencies?
- Do roles vary according to the type of emergency?
- What training have response personnel experienced?
- What training is necessary?

Assessing Capability to Conduct an Exercise

To design an exercise that simulates a real emergency, you must know what responses are planned and assess what capabilities are needed to meet those responses. Then, you must ask if your jurisdiction is at a point where it can conduct an exercise.

Before launching an exercise, it's important to find out if you have the resources—skills, funding, personnel, time, facilities, and support. Deficiencies in any of these areas must be considered in the design process. When sheer enthusiasm leads the jurisdiction to want to conduct a full-scale exercise, sometimes the capability assessment will show that the jurisdiction is ready only for a tabletop exercise.

Questions to Ask About Capability and Resources

The following questions, though very general, can help you assess your jurisdiction's level of capability. For example, you may find that before you consider planning an exercise, you will need to develop, support, and train people.

- When was your jurisdiction's last exercise?
- What exercise experience is available in your jurisdiction? What is your own experience? What is the experience of your staff?
- How much preparation time can you reasonably expect to have allocated to developing an exercise?
- How much time can people devote to developing an exercise?
- What skills can those people provide?
- What physical facilities do you use when you conduct an emergency operation? Will they be available for an exercise?
- What communication facilities and systems do you use in a real emergency? Will they be available for an exercise?
- What attitudes do you expect of the chief elected official and emergency services providers or other leaders toward the exercise?

Addressing Liabilities and Costs

Liabilities and costs should be addressed early. A problem inherent in many exercises, particularly drills and full-scale exercises, is the possibility of injury or damage to equipment. *Before planning any exercise, check your jurisdiction's insurance coverage.* Costs (both apparent and hidden) are incurred at every stage of exercise development. Plan ahead to be sure that your jurisdiction has the needed resources. Some cost considerations are outlined next.

Cost Considerations

Plan for a wide variety of costs. The following are a few examples; actual costs will vary with the exercise and depending on your jurisdiction's policies.

- Staff salaries
- Contract services
- Equipment and materials
- Fuel to run equipment and transport volunteers
- Miscellaneous items

Ask some key questions to avoid committing more resources than are available. For example:

- Will reimbursement for overtime be required if the exercise takes place on a weekend or an evening?
- If the exercise supports a hospital certification exercise, who will cover the costs?

Employees should have their emergency management responsibilities reflected in their job titles.

Time for participation in training, planning, and exercising should be set aside for each employee who has an emergency management responsibility.

Costs for routine participation in exercises should be recognized by jurisdiction officials.

It is important to determine the cost of an exercise to the jurisdiction. It is equally important to determine the jurisdiction's readiness for the exercise process. See Figure 8.5.

Gaining Support

From the beginning of the exercise design process, it is important to establish authority for conducting the exercise. This means gaining the support of the highest possible official in your jurisdiction.

Even if the chief elected official (CEO) does not participate in the exercise, his or her authority and support are essential. The top official can get nearly instant and complete cooperation from those who will participate in the exercise. Without that authority, it can be very difficult to put on an exercise involving departments, organizations, or agencies over which you personally have no authority.

Gaining support from the CEO is not always easy, but the following approaches will help:

Self-Assessment: Resources and Costs

Plans

How familiar are you with the emergency plans, policies, and procedures of your jurisdiction?

☐ Very familiar

☐ Only general familiarity

☐ Familiar with only a portion

☐ Need to review plans, policies, and procedures thoroughly

Time

How far in advance would your jurisdiction realistically have to schedule to plan and design each of the following exercise activities effectively?

• Orientation _____

• Drill _____

• Tabletop exercise _____

• Functional exercise _____

• Full-scale exercise _____

How much preparation time can reasonably be allocated to developing an exercise?

• Actual person days:

• Elapsed time to exercise:

Self-Assessment: Resources and Costs

Experience

When was your jurisdiction's last exercise?

What is your previous experience with exercises? (Check all that apply.)

Orientation:	☐ Presenter	☐ Participant		
Drill:	☐ Controller	☐ Participant		
Tabletop exercise:	☐ Facilitator	☐ Participant		
Functional exercise:	☐ Controller	☐ Simulator	☐ Player	☐ Evaluator
Full-scale exercise:	☐ Controller	☐ Responder	☐ Evaluator	☐ Victim

Took part in postexercise debrief

Helped write an evaluation report

What other exercise-related experience is available in your jurisdiction?

Figure 8.5   Getting ready for exercise design.   *Continued*

Facilities

What physical facilities do you use when conducting an emergency operation?

Will they be required for this exercise?   Yes ☐   No ☐
Will they be available for this exercise?   Yes ☐   No ☐

Self-Assessment: Resources and Costs

Communications: What communications facilities and systems do you use in a real emergency?

Will they be required for this exercise?   Yes ☐   No ☐
Will they be available for this exercise?   Yes ☐   No ☐

Barriers: Are there any resource barriers that need to be overcome to carry out this exercise?   Yes ☐   No ☐

If so, what are the barriers and how can they be overcome?

Self-Assessment: Resources and Costs

Costs

What types of costs might be incurred for these exercises in your jurisdiction? (Do not list exact figures—just types of expenses, such as wages and salaries, transportation, etc.)

For an orientation:

For a drill:

For a tabletop exercise:

For a functional exercise:

For a full-scale exercise:

Are there ways that different organizations or agencies can reduce costs (e.g., by combining exercises, cost-sharing, resource-sharing) and still fulfill program requirements? Explain.

Figure 8.5 (*Continued*)   Getting ready for exercise design.

- Gain support for the entire exercise program—Build a comprehensive, progressive exercise program. The CEO will be more receptive to an exercise that is part of a proven, consistent, and goal-oriented program than to an isolated exercise.
- Protect the jurisdiction—Make a conscientious effort to protect the jurisdiction from lawsuits.
- Sell the process—Your needs assessment, capability analysis, purpose statement, and objectives are important building blocks for the exercise. Beyond that, they provide a valuable tool for selling the idea professionally—first to your boss and later to the CEO in your jurisdiction.

Presenting the Concept

Having the results of your early preparation activities in hand will add to your credibility when presenting the exercise concept to those in authority. Your presentation should include brief explanations of:

- The need for the exercise
- Organizational capability (experience, personnel, costs)
- The type of exercise
- The scope of the exercise
- The purpose of the exercise

Announce the Exercise

Broad support for the exercise may be gained in some instances by sending out an announcement—often in the form of an exercise directive. It should come from the CEO, but you need to be prepared to write it. The directive serves the purpose of authorizing you to conduct the exercise and giving you the clout you need to gain support from others.

Write the Exercise Directive

The exercise directive will closely resemble the purpose statement. The directive should contain the:

- Purpose
- List of participating agencies
- Personnel responsible for designing the exercise
- Approximate exercise date
- Point of contact for additional information

Notice that the nature of the emergency and the location of the exercise are not revealed.

February 24, 20XX

TO: All Agency Directors
FROM: Mike Fagel
           Chief Administrative Officer
SUBJECT: Emergency Exercise

A simulated emergency exercise involving a terrorist incident has been scheduled for some time during the week of January 12–18, 20xx.

The purpose of the proposed exercise is to improve the following emergency operations:

Rapid assessment.
Notification and alert.
Scene isolation and perimeter control.
Mass casualty triage.

It is important that your agency participate in this exercise. We encourage involvement at the highest level.

I believe we all realize the importance of emergency exercises as a means to community preparedness. I fully support this exercise and intend to join with you in participating.

The Emergency Management Office will be coordinating the exercise. They will be contacting you to make necessary arrangements for the development and conduct of the exercise. For purposes of realism and interest, details of the exercise situation will not be made known prior to the exercise.

For further information, call Planning at EXT 1234.

Figure 8.6   Sample exercise directive.

Regarding dates: Totally unannounced exercises are not recommended. However, whether you specify the exact date will depend on the degree of surprise that you intend. At a minimum, a range of dates should be given. In selecting the exercise date, check the community calendar to avoid conflicting with a major event. Figure 8.6 is an example of an exercise directive.

Assembling a Design Team

Planning an exercise requires a multitude of tasks, from designing the exercise to arranging detailed administrative matters. These tasks require the effort of a dedicated team and team leader.

Exercise Design Team Leader

The exercise design team leader is responsible for the exercise throughout the entire development process—and for managing all administrative and logistical matters.

In an exercise involving multiple jurisdictions, the team leader may need assistants or a liaison from each entity to help coordinate the many details.

Who Should Lead?

Because the leader is so important, it is essential that this person be experienced and capable. The exercise design team leader should be someone who:

- Can devote a considerable amount of time throughout the exercise cycle.
- Is familiar with the emergency plan and has a sound understanding of the response organizations that will be participating.
- Is not a key operational member of one of the participating jurisdictions. All key members should participate in the exercise, and they might not be able to participate fully if they are involved in the design process.

What about the Emergency Manager?

Generally speaking, the person with chief responsibility for managing emergency events should be a player in the exercise. Therefore, instead of heading the design team, if you are the emergency manager for your jurisdiction, you should assign someone else to develop and conduct the exercise. The reason for this is that, before the exercise, the emergency manager should be given only the information that other players receive—but no more.

When staffing does not permit the emergency manager to assign someone, he or she will have to play a dual role:

- Assume responsibility for developing the exercise and gather support from other agencies
- Participate in the exercise, but on a limited basis

Despite being familiar with the exercise design, the emergency manager can still take a number of actions without compromising the realism.

Exercise Design Team

Exercise design is a complex task that should not be done by a single individual. The exercise design team assists the team leader in developing exercise content and procedures.

Design Team Responsibilities

Design team members:

- Determine the exercise objectives
- Tailor the scenario
- Develop the sequence of events and associated messages
- Assist in the development and distribution of preexercise materials
- Help conduct preexercise training

Ultimately, team members will be good candidates to act as simulators or controllers in a functional exercise.

Design Team Candidates

Ideally, the team should include a representative from each of the participating agencies and from participating jurisdictions in a multijurisdictional exercise. If this results in an unwieldy team, select a small core team that can draw on other members as needed.

Backgrounds

Selecting team members with varied backgrounds helps with coordination and stimulates creativity. Some additional technical administrative support may be required for typing, printing, and other mechanics involved in the preparation of materials.

Organizing the Design Team

Design teams are so varied—in number of members, team makeup, available skills, political climate, scope of the exercise to be developed, and many other factors—that there is no single approach to organizing a design team.

Generally, the team leader should use the kinds of teamwork strategies that make any team successful. Team leadership is too large a topic to discuss in this chapter, but a few suggestions follow.

Strategies for Organizing the Team

- Establish clear goals.
- Agree on a plan of action, with specific responsibilities and due dates, to meet established goals.
- Reach consensus on a realistic schedule for completing milestones.
- Meet regularly to monitor progress.
- Work together to share expertise and resources to get the job done.
- Use team interaction to learn more about one another's agencies or jurisdictions. The more you know about other stakeholders, the better you will able to work together to manage emergencies when they arise.

- Keep lines of communication open about new ideas, potential pitfalls, successes, problems, and needs. Shared information and creative problem solving are huge assets in emergency management.
- Use job aids to make the schedule, responsibilities, and progress visible and to keep everyone on the same page.

Some examples of job aids are shown on the following pages. Remember, these are only examples. Any job aid that you create should be adapted to meet the needs of your exercise team.

Checklists. A simple to-do list can be used to provide an overview of the process and ensure that all main tasks are completed (see Figure 8.7).

Activities schedule. For relatively simple exercises, a basic schedule may be used to show major activities of the team and completion deadlines for each (see Figure 8.8).

A more detailed plan is needed for a major exercise, which must be planned with the thoroughness of any major organizational effort. A Gantt chart, as shown in Figure 8.9, is useful as a scheduling tool for such detailed planning. A Gantt chart displays time across the top and a sequence of tasks down the left-hand side. Time can be given in days, weeks, or months. The duration of the time devoted to each activity is represented by bars extending across the timelines. Figure 8.9 shows task groupings (with supporting subtasks scheduled) and staff allocations indicated by initials in the left column.

Exercise Documents

Four major documents are developed during the exercise design process:

1. Exercise plan
2. Control plan
3. Evaluation plan
4. Player handbook

These documents are basically handbooks for specific audiences and serve as useful tools during exercise development, conduct, and evaluation.

Exercise Plan

The exercise plan contains information that everyone needs, and it serves many purposes. For example, it can be used:

- To provide general information about the exercise for everyone concerned, including exercise overview, parameters, and timelines

Mission	Scenario
☐ Needs Assessment	☐ Narrative
☐ Scope	☐ Major/Detailed Events
☐ Statement of Purpose	☐ Expected Actions
☐ Objectives	☐ Messages
Personnel	Logistics
☐ Design Team	☐ Safety
☐ Controller or Facilitator	☐ Scheduling
☐ Players	☐ Rooms/Location
☐ Simulators	☐ Equipment
☐ Evaluators	☐ Communications
☐ Management	Phones
Safety	Radio
Observers	Computers
Information	☐ Enhancements
☐ Directives	Maps
☐ Media	Charts
☐ Public Announcements	Other:
☐ Invitations	Evaluation
☐ Community Support	☐ Methodology
☐ Management Support	☐ Locations
☐ Timeline Requirements	☐ Evaluation Forms
Training/Briefings	☐ Postexercise Debrief
☐ Train Simulators, Evaluators, Controllers	After-Action Documentation/ Recommendations
☐ Players' Preexercise Briefing	☐ Evaluation Meeting
	☐ Evaluation Report
	☐ Follow-Up Ideas for Next Exercise

Figure 8.7 Sample exercise development checklist.

Deadline for Completion	Leader Activities	Team Activities
3 months prior	1. Hold initial planning meeting	
2½ months prior	1. Brief government officials 2. Arrange for facilities 3. Determine simulation structure 4. Convene and brief design team	1. Attend team briefing
2 months prior	1. Review and finalize scenario	1. Develop/review exercise procedures 2. Arrange simulation 3. Arrange participation 4. Review exercise scenario
1½ months prior	1. Obtain exercise materials 2. Prepare ideas for scripted messages	1. Prepare participant information packet 2. Prepare operational data
1 month prior	1. Review messages with team	1. Review messages with leader 2. Review evaluation forms 3. Print forms 4. Prepare scripted messages
3 weeks prior	1. Prepare briefing for participants	
2 weeks prior		1. Integrate messages into time schedule 2. Develop training sessions
1 week prior	1. Prepare exercise facility	
2–4 days prior	1. Conduct training session 2. Train supervisors	1. Assist in training sessions
Day of Exercise	1. Conduct participant briefing 2. Perform preexercise check 3. Supervise the exercise	1. Assist with preexercise check
1 week after	1. Help prepare draft of final report	1. Review final report and make suggestions
2 weeks after	1. Revise and submit report	
3 weeks after	1. Submit recommendations	

Figure 8.8   Sample activities schedule.

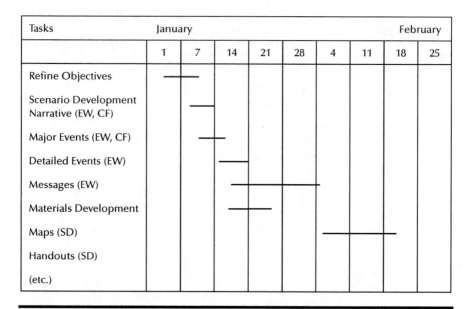

Tasks	January					February			
	1	7	14	21	28	4	11	18	25
Refine Objectives	—								
Scenario Development Narrative (EW, CF)		—							
Major Events (EW, CF)		—							
Detailed Events (EW)			—						
Messages (EW)				—					
Materials Development			—						
Maps (SD)						—			
Handouts (SD)									
(etc.)									

Figure 8.9   Sample Gantt chart.

- As a guide for developers and participants
- To assist participants in enhancing exercise play
- As a promotional tool by the exercise director

Control Plan

The control plan is for controllers and simulators. It is not made available to players. It provides information about controller and simulator requirements and explains the exercise concept as it relates to controllers and simulators. This plan:

- Establishes the basis for control and simulation of the exercise
- Explains the procedures, responsibilities, assignments, and support for exercise control and simulation
- Defines the communications, logistics, and administrative structure needed to support control and simulation during the exercise

Evaluation Plan

The evaluation plan provides exercise evaluators, as well as controllers and simulators, with guidance concerning exercise evaluation procedures, responsibilities, and support. This plan:

- Explains the exercise concept as it relates to the evaluation process

- Establishes the basis for evaluation
- Defines the communications, logistics, and administrative structure needed to support evaluation before, during, and after the exercise

Player Handbook

The player handbook provides exercise players with information needed to participate effectively in the exercise. This information is also discussed at the player briefings that are conducted just prior to the start of the exercise. Specific information included in these documents is presented in Table 8.8.

Exercise design is much like scripting a play to make sure that all of the players perform the correct actions and make the right decisions at the appropriate time. Tabletop, functional, and full-scale exercises are based on a design process that includes the eight steps addressed earlier in the chapter.

Step 1: Assess Needs

Some jurisdictions plan exercises as a response to pressure. For example, someone may suddenly decide that it is necessary to do a full-scale exercise as a response to some dramatic disaster. Such a hasty decision will usually result in failure and embarrassment. The best way to determine where you need an exercise—and what kind of exercise is needed—is to study your situation systematically.

Conducting a needs assessment will help you define the problems, establish the reasons to do an exercise, and identify functions to be exercised.

Begin with Your Plan

A needs assessment should begin with a review of the emergency plan and should address:

- Hazards—the risks that you are most likely to face and the priority levels of those hazards
- Area(s) most vulnerable
- Functions most in need of rehearsal
- Potential participants (agencies, organizations, departments, personnel)
- Past exercises
- Exercise requirements
- Capabilities

If you have addressed your jurisdiction's exercise needs when planning a comprehensive exercise program, you have a good base upon which to plan your next steps. Consulting and updating that assessment will be a crucial step whenever a new exercise is considered for development.

Table 8.8 Contents of Exercise Design Documents

Exercise Plan	Control Plan	Evaluation Plan	Player Handbook
• Exercise type and purpose	• Exercise concept	• Exercise concept	• Exercise scope (concept of play, assumptions, artificialities, and simulations)
• Scenario narrative	• Preexercise player activity	• Preexercise player activity	• Scenario narrative
• Scope	• Assumptions, artificialities, and simulations	• Assumptions, artificialities, and simulations	• Player procedures and responsibilities
• References	• Concept for management, control, and simulation	• Concept for evaluation management	• Safety and security
• Assumptions, artificialities, and simulations	• Control team staffing	• Evaluation team staffing	• Communications
• Objectives	• Control team training	• Evaluation team training	• Reporting
• Concept of operations	• Control team staff responsibilities	• Evaluation team staff responsibilities	• Administrative systems
• Exercise management structure and responsibilities	• Control team procedures	• Evaluation team procedures	• Recommended preexercise training events
• Safety and security	• Communications, logistics, administrative, and other support	• Support for the evaluation team	• Schedule of player exercise briefings
• Administration and logistics			• Provisions to review community plans, policies, and procedures
• Exercise management job aids			• EOC (or other operating center) procedures

Lessons Learned

In doing a needs assessment for a single exercise, an obvious starting point is the evaluations of past exercises:

- Who participated and who did not?
- To what extent were the objectives achieved?
- What lessons were learned?
- What problems were revealed, and what is needed to resolve them?
- What improvements were made following past exercises, and have they been tested?

Needs Assessment Results

In summary, your needs assessment should reveal the following types of issues if they exist:

- Primary and secondary hazards that the jurisdiction faces
- Problems that need to be resolved
- Problems that recur
- Skills that need to be practiced
- Functions that are weak
- Improvements implemented after previous exercises, which now need to be tested
- New facilities, personnel, or equipment that have not been included in an exercise
- Weaknesses (such as gaps, conflicting policies, or vague procedures) in the emergency plan
- The need for role clarification
- The need for a certain type of exercise

Step 2: Define the Scope

A needs assessment may reveal a wide array of concerns. Clearly you can't design an exercise that practices:

- All functions ...
- In the context of all hazards ...
- Using all agencies, organizations, or departments ...
- In all exercise formats ...
- Employing all resources ...

You will need to set priorities and make choices. It is important that the scope be clearly and narrowly defined.

How Scope Is Determined

Many factors influence which areas of concern will be included in an exercise and which will not. Sometimes, one decision will influence another. Other factors that help to define the scope of an exercise include:

- Expense
- Availability
- Seriousness of the problem
- Capability of the exercise to address the problem
- Skills and experience of the designers
- Length of the exercise

What Does Scope Include?

There are five key elements of scope: type of emergency, location, functions, participants, and exercise type.

1. Type of emergency—An exercise is usually limited to one major event, although others—especially secondary events—might develop as the scenario unfolds. Hazards may be chosen for several reasons, including:
   - The emergencies that will generate the types of actions that need to be practiced.
   - The highest priority hazards that the organization faces.
   - The hazards that haven't been exercised recently.
   - Problems that have just recently developed.
2. Location—Identify the location where the simulated event will occur. For tabletop and functional exercises, select a place where the hazard could typically occur. For a full-scale exercise, traffic problems or safety issues may make it necessary to compromise on an area similar to the ideal location.
3. Functions—List the operations that the participants will practice. Be sure that the procedures within a certain function are clear and narrowly defined.
   Example: To exercise a community's alert warning system, the following actions might be part of a response function:
   - Notify the warning agency.
   - Turn on sirens.
   - Notify fire or police to use loud speakers in area.
   - Notify Emergency Alert System (EAS) to interrupt programming with message.
4. Participants—After the most important functions or needs have been identified, you can narrow the list of participating agencies and individuals to those who are required to carry out the actions.
   Ask yourself:

- Which organizations need to be involved to carry out the function(s) being tested?
- Which representatives from the identified organizations should be there?

For example, in an EOC or other operations center, you would typically want policy makers, coordinators, and operations personnel. In an incident command post, you would most likely want personnel who are knowledgeable in field operations and have some on-scene decision-making authority.

5. Exercise type—Finally, a decision must be made on the type of exercise to be performed.

Ask yourself:
- What exercises are most needed?
- What experience have personnel had with the various types of exercises?
- What stress level do we want?
- What types of exercises are mandated by regulatory requirements?

After these issues have been settled, it's time to formulate them into a statement of purpose.

Step 3: Write a Statement of Purpose

As mentioned earlier in the chapter, the statement of purpose is a broad statement of the exercise goal. The purpose statement:

- Governs the selection of the objectives, which in turn govern subsequent steps
- Clarifies for the CEO and potential participants why the exercise is being conducted
- Is useful in communicating plans to the media and community leaders

Developing the Purpose Statement

A purpose statement is easily constructed. One approach is simply to incorporate the scope decisions (type of emergency, locations, functions to be tested, etc.) into a single sentence. A date is then added. Two sample formats are shown in Figure 8.10 and Figure 8.11.

The exercise directive, as discussed earlier in the chapter, is a memo from the CEO in your jurisdiction that is sent to agencies or departments whose support you need. The directive is essentially a restatement of the purpose statement.

When using the statement of purpose as the basis for the directive, the following information should be added:

- Contact person and telephone number
- Hours that the exercise will be conducted
- Exercise location (may be omitted if you wish to retain an element of surprise)

The purpose of the proposed emergency management exercise is to improve the following emergency *operations*:

Flood stage monitoring
Evacuation warning
Relocation of school children
Shelter management

by involving the following *agencies*:

Emergency Management
Fire Department
Public Works
Health Department
Red Cross
Public Schools

in a functional exercise simulating a flash flood

at Planter's Street Bridge to Route I-740 on April 2.

Figure 8.10   Statement of purpose—Sample 1.

The purpose of the proposed emergency management exercise is to coordinate the activities of city and county government, volunteer organizations, and private industry in their response to a major incident; to provide training to staff; to test and evaluate the alert and warning, evacuation and shelter/mass care annexes; and to enhance interagency coordination and cooperation by involving the following department or agency heads:

County Commissioner or Chief Administrative Officer	Justice County
Mayor	City of Liberty
Emergency Manager	City of Liberty
Emergency Manager	Justice County
Fire Chief	Justice County
Law Enforcement	Justice County Sheriff
PIO	Liberty City Gazette
HazMat Team Liaison	Representative
Chemical Expert	Justice County Team #3
Poison Control Center	Arrow Chemical Company
The American Red Cross	Dr. Smith, Disaster Director
Liberty City Hospital	Emergency Room Director

These entities will be tested on July 15, in a simulated exercise involving a hazardous materials transportation accident at SW Mail Road near SW Johnston Boulevard, approximately 300 yards from the Liberty City Hospital.

Figure 8.11   Statement of purpose—Sample 2.

Exercise:
Scope:
Type of Emergency:
Location:
Functions:
Organizations and Personnel:
Exercise Type:
Statement of Purpose:

Figure 8.12   Define exercise scope and purpose.

Figure 8.12 is useful in helping define the exercise scope and purpose.

Step 4: Define Objectives

Early in the development of an exercise, you must decide what the exercise is intended to accomplish. These outcomes, or objectives, must be specified clearly. An objective is a description of the performance that you expect from participants to demonstrate competence. Objectives go hand in hand with the purpose statement but are more specific and performance based.

Why Define Objectives?

Objectives are essential during the four stages of the exercise process:

1. Design process—Objectives are the pivot point in the design process. The needs assessment, scope, and purpose statement lead to the formulation of objectives. The success of later actions and decisions begins with carefully written objectives. The narrative, the major and detailed events, expected actions, and messages are all based on objectives. In one sense, the objectives can be thought of as general statements of expected actions.
2. Exercise conduct—During the exercise itself, elements of the exercise should be conducted according to the objectives to make sure that it stays on track.
3. Evaluation—Writing objectives is the beginning of the exercise evaluation process. During the exercise, observers use the objectives to evaluate performance. After the exercise, the evaluation report is based upon these objectives. The process of identifying evaluation criteria takes place at the time the objectives are written.
4. Follow-up—During the follow-up period, participants retrain, plan, and practice to address objectives that were not fulfilled.

How Are Objectives Determined?

Many objectives become evident at the time of the needs assessment, when designers identify problem areas. These needs can usually be translated into a statement of objectives. Objectives are also arrived at by breaking down a purpose statement into its logical components.

Example: Suppose your last exercise showed weaknesses in alert and notification—specifically, a failure on the part of the EOC to analyze and implement calldown procedures. One of the resulting objectives would be to verify that the EOC is now able to notify the proper agencies according to the plan.

How Many Objectives?

There can be as few as two or three objectives in a small exercise, or as many as a hundred in a large national exercise including many federal, state, and local jurisdictions. For an average exercise, ten or fewer objectives are recommended.

In larger exercises, each participating organization should be responsible for developing its own specific objectives, which are then incorporated into one exercise package by the design team.

What Makes a Good Objective?

The main thing to remember about objectives is that they must be clear, concise, and focused on participant performance. They should contain:

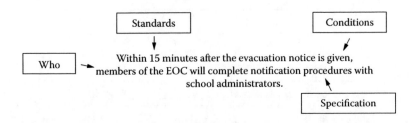

Figure 8.13 An example of a good objective.

- An action, stated in observable terms
- The conditions under which the action will be performed
- Standards (or level) of performance

In other words, an objective should state *who* should do *what* under *what conditions* according to *what standards*. What each arrow points to in Figure 8.13 illustrates the parts of an objective.

Writing SMART Objectives

A useful guideline for writing objectives is the SMART system, which is described in Table 8.9. The system is simple to use and easy to remember.

Table 8.9 SMART Guidelines for Useful Objectives

Simple	A good objective is simply and clearly phrased. It is brief and easy to understand.
Measurable	The objective should set the level of performance, so that results are *observable*, and you can tell when the objective has been achieved. This doesn't mean that you have to set a quantifiable standard. It just means that people can agree on whether they succeeded.
Achievable	The objective should not be too difficult to achieve. For example, achieving it should be within the resources that the organization is able to commit to an exercise.
Realistic	The objective should present a realistic expectation for the situation. Even though an objective might be achievable, it might not be realistic for the exercise.
Task oriented	The objective should focus on a behavior or procedure. With respect to exercise design, each objective should focus on an individual emergency function.

Display	Yes	No
Status boards available in facility	☐	☐
Status boards used	☐	☐
Status boards kept updated by _____	☐	☐
Maps available	☐	☐
Maps up to date	☐	☐

Figure 8.14 Examples of specific points of review.

Points of Review

Another way to ensure that the objective will provide useful measures is to include specific points of review—very specific items to be observed by an evaluator. For example, the following objective is a little too vague to guide an evaluation team.

Objective: Demonstrate the adequacy of displays and other materials to support emergency operations.

If the objective is supported by points of review similar to Figure 8.14, it will be adequate. Also, Figure 8.15 offers some helpful action verbs and concrete terms to provide specificity to an exercise and review.

Most objectives that are written for exercises describe performance—that is, the actions that a person or organization will carry out. Occasionally, the objective describes the understanding of a concept or change in attitude. When writing objectives, remember to use the SMART system and include:

- Action, stated in observable terms
- Conditions
- Standards

Step 5: Compose a Narrative

An exercise is a scenario that simulates an emergency. Part of this scenario is the narrative, which is a brief description of the events that have occurred up to the minute the exercise begins. The narrative has two important functions. First, it *sets the mood* for the exercise. Participants need to be motivated to participate. The narrative captures their attention and makes them want to go on. Second, the narrative *sets the stage for later action* by providing information that the participants will need during the exercise.

Helpful Hints: Word Choice

Use concrete words. One way to avoid vagueness is to use concrete words. Pay particular attention to the verb that describes participant performance.

Avoid vague verbs, such as:

- Know
- Understand
- Appreciate
- Show the ability to
- Be aware of

Use action words, such as:

- Assess
- Clarify
- Define
- Determine
- Demonstrate
- Establish
- Evaluate
- Examine
- Explain
- Identify
- Inspect
- List
- Notify
- Operate
- Prepare
- Record
- Report
- Show
- Test

Figure 8.15 Helpful action verbs and terminology to use.

Characteristics of a Narrative

A good narrative:

- Is usually one to five paragraphs long
- Is very specific
- Is phrased in present tense
- Is written in short sentences to lend immediacy and tension
- May develop the situation chronologically (event with warning time)
- May emphasize the emergency environment

For an emergency with warning time (e.g., a hurricane), the narrative often outlines the developing situation chronologically.

For an unexpected event (e.g., a chemical spill or terrorist bomb attack), the narrative may be shorter. Or, it may devote more detail to the environment of the emergency (e.g., a nearby school, other chemicals stored, rush hour approaching) to create intensity of feeling.

Hints: Outlining a Narrative — You can outline a narrative by jotting down short responses (one or two words) to the following questions:

- What event?

- How fast, strong, deep, dangerous?
- How did you find out?
- What response has been made?
- What damage has been reported?
- What is the sequence of events?
- What time?
- Was there advance warning?
- Where does it take place?
- What are the relevant weather conditions?
- What other factors would influence emergency procedures?
- What is predicted for the future?

Then, when you're ready to write the text of the narrative, just take each of the key words and turn it into a brief scenario.

What follows are two sample narratives. As you read the narratives, try to notice where the aforementioned questions are answered.

Sample Narrative: Hurricane

The National Weather Service's National Hurricane Center issues news on the formation of a storm off the southern U.S. coast that appears to have hurricane potential. Tropical Storm Anne is upgraded to Hurricane Anne, and NWS issues a hurricane watch for a three-state area along the U.S. coast. Wind velocity and northwest movement over the last day have decreased, but an overnight change in direction to a steady northwest line calls for an immediate hurricane warning for five coastal counties of the state. Winds of 120 mph are predicted during the incoming tide, with high water expected to reach twelve to fifteen feet over high tide. Threatened areas include low-lying, newly developed resort areas. A heavy influx of visiting weekend campers has been advised to evacuate the area. Access bridges to barrier islands are narrow and could become impassable with fifteen-foot water heights.

Hurricane Anne, considered a very dangerous hurricane with high winds and an accompanying storm surge, will hit the coastal communities along Stevens Bay and farther inland, a population area of between 5,000 and 25,000.

Following the issuance of the hurricane watch, emergency service personnel notified elected officials and agency heads within the watch area. News media were also alerted and encouraged to broadcast the notice. When the warning of landfall within twenty-four hours was given, the emergency manager placed her staff on alert but did not activate the EOC. She has asked all appropriate emergency service personnel to meet at 0730, approximately four hours after the warning was given. On its present course, the hurricane will make landfall at approximately 2330. Flood stage from rising tides and tidal surge could, however, impact bridges by 1600. All appropriate staff and emergency personnel are now gathered in the EOC.

Sample Narrative: Air Crash

A Boeing 747, en route from Panama to San Francisco, is experiencing in-flight engine problems and will have to make an emergency landing. Plans have been made to land at a large airport 200 miles north. However, the latest communication with the pilot indicates that the plane has lost engine power and is losing altitude too quickly to reach the large airport. Even though your city airport is too small to handle a 747, you are the only hope for the 350 passengers and 10 crew members.

Conditions at your airport are clear, and the surrounding area is dry. A hot, dry wind is blowing from the north.

The main runway lies along a relatively unpopulated suburban area. However, the likelihood of the pilots being able to control the huge plane and stay within the landing space is slim. The approach passes over populated suburban housing developments.

The airport control tower alerts its own crash/fire rescue units and requests that the local emergency services provide backup assistance in fire, police, medical, welfare, and search and rescue capabilities.

Step 6: Write Major and Detailed Events

Developing an exercise scenario is much like writing a play. In developing a play, the playwright organizes events into acts and scenes. Similarly, an exercise designer organizes events into major and detailed events. Major and detailed events are occurrences—large or small—that take place after and as a result of the emergency described in the narrative. It may be helpful to think of them as problems requiring a realistic action that will meet exercise objectives.

The goal in developing events is to provide a structure that will link the simulated event to the actions that you want people to take and provide unity to the exercise. Without the overall organization provided by major events, the exercise could dissolve into random actions.

Careful scripting is very important if you are going to produce a convincing, unified scenario rather than a series of unrelated, miscellaneous occurrences. It is also necessary for creating an exercise that is governed by objectives.

Developing Major Events

Major events are big problems resulting from the emergency. They should be likely events, based on case studies or operational plans, that call for realistic action.

Usually, the best way to arrive at a list of major events is to take it in two stages:

1. First, identify several major occurrences—the high points in a sequence—that might follow the narrative events.
2. Second, decide which of these events might generate situations that would test the objectives. Then, concentrate on those which best support the objectives.

The major events in the following example show that most events require actions from one or more departments or agencies.

Sample Major Events Sequence for Air Crash Scenario

1. Fuselage breaks apart and hits buildings below.
2. Jet fuel ignites several homes in the area.
3. About 60 survivors are thought to be trapped in the front section of the aircraft.
4. Several bystanders are injured on the ground.
5. A crowd convenes around the crash site.
6. Family members of victims begin to gather at the crash site.
7. Estimates of passenger casualties rise between 200 and 220 deaths and at least 70 severe-burn victims.

Because your goal is to develop an exercise that will test certain functions and departments, the major events should be developed from your purpose statement.

Example:

■ Event #1 tests damage assessment, and command and control.
■ Event #2 tests deployment of fire, police, and medical services.
■ Event #3 tests search and rescue and medical personnel procedures.

Developing Detailed Events

Detailed events are specific problem situations to which personnel must respond. Each detailed event should be designed to prompt one or more expected actions from one or more organizations that are participating in the exercise. When the design task is small, it may not be necessary to distinguish between major and detailed events. For functional exercises, however, it is easier to write messages if you can base them on a list of detailed events.

There are several ways to develop detailed events. For example, you can:

- Plan the detailed events and expected actions at the same time.
- Work backward—first identifying an action that you want players to perform, then listing a problem (a detailed event related to a major event) that would motivate the action.
- Make a list of specific problems that are likely to occur in connection with each major event, then identify actions that would be expected as a result.

Whatever method you use, the result should be a list of specific events that are closely linked with actions that you want the participants to perform.

Step 7: List Expected Actions

Expected actions are the actions or decisions that you want participants to carry out to demonstrate competence. It is necessary to identify expected actions to:

- Write messages—Because the point of the exercise is to get the participants to think and react in certain ways, the script must be carefully developed to ensure that the messages get the planned results. Your list of expected actions will enable you to write effective messages.
- Determine what should be evaluated—The exercise evaluation will focus on whether the participants respond appropriately in an emergency. The list of actions will become the core of that evaluation.

Types of Actions

There are four types of actions that the participants may carry out:

1. Verification—Gather or verify information
2. Consideration—Consider information, discuss among players, negotiate, consult plan
3. Deferral—Defer action to later, put action on priority list
4. Decision—Deploy or deny resources

You will know what actions are appropriate in response to a given event by referring to the emergency plan.

Relationship to Objectives

Expected actions are closely tied to objectives. Objectives state general desired actions. Expected actions are a breakdown of objectives—the actions that would be taken by an organization or an individual to meet the objective. Table 8.10 illustrates this relationship.

Generally speaking, every detailed event results in one or more expected actions from various organizations. When you prepare a list of expected actions:

- List only those that involve the participating departments or agencies—those identified in the exercise scope and the statement of purpose.
- List actions for all exercise participants. It is not necessary that each detailed event generate responses from all participants.

Figure 8.16 will help you write major and detailed events. For each detailed event that you just listed, identify the expected action, the organization responsible, and the objectives that it addresses (see Figure 8.17).

Table 8.10 An Example of Objective and Expected Actions

Function	Coordination and communication among the airport and the jurisdiction's emergency systems.
Objective	Upon notification that a crash is imminent, response units will stage within three minutes, according to SOPs.
Event	Landing of disabled aircraft is imminent.
Expected actions	Airport Control Tower: • Notify police, fire, medical personnel to proceed to airport. • Alert hospitals of potential mass casualty incident. Dispatch Center: • Alert police, fire, and medical supervisors. Hospital: • Notify other medical facilities as appropriate. Crash Fire Rescue: • Initiate incident command system. • Notify dispatch of command post and staging locations.

Major Event #1:
Detailed Events: 1. 2.
Major Event #2:
Detailed Events: 1. 2.

Figure 8.16   Writing major and detailed events.

Step 8: Prepare Messages

Messages are used to communicate detailed events to exercise participants. One message may be used to represent an event, or several messages may be needed to notify the participants of the event. Messages serve one purpose: to evoke a response—that is, to make decisions and take actions that meet the exercise objectives.

In a full-scale exercise, the controller may input prescripted messages into the action. Participants receiving the messages would then make decisions or take action as they would in a real emergency.

Transmitting Messages

Messages can be transmitted in various ways, including:

- Landline/cellular telephone
- Radio
- In person
- Written note
- Fax

When transmitting messages in functional or full-scale exercises, it is important to make every effort to use the method of transmission that would be likely in an actual emergency.

Detailed Event	Expected Action	Organization

Figure 8.17　Expected actions planning worksheet.

Relationship to Expected Actions

Messages have a direct relationship to expected actions. Each message is designed to generate one or more expected actions. Table 8.11 lists some ideas for messages that could achieve the expected actions.

Message Variables

Messages—whether simple or complex—have four main variables, which are listed next. Not all variables will be explicitly stated in every message, but they should be kept in mind as you write because these variables form the classic definition of communication: *Who* sends *what* to *whom*, with what *effect*.

Table 8.11   An Example of Objective and Expected Actions

Function	Coordination and communication among the airport and the jurisdiction's emergency systems.
Objective	Upon notification that a crash is imminent, response units will stage within three minutes, according to SOPs.
Event	Landing of disabled aircraft is imminent.
Expected Actions	Airport Control Tower: • Notify police, fire, medical personnel to proceed to airport. • Alert hospitals of potential mass casualty incident. Dispatch Center: • Alert police, fire, and medical supervisors. Hospital: • Notify other medical facilities as appropriate. Crash Fire Rescue: • Initiate incident command system. • Notify dispatch of command post and staging locations.
Possible messages	• Radio call from plane to tower. • Tower calls police, fire, rescue. • Plane requests runway be designated. • Call from hospital requesting information. • Calls to dispatch from media. • Degrading radio communications with plane. • Pilot feels major vibrations/noise on the plane.

■ Message source (who): Who sends the message (must be a credible source).
■ Transmission method (how): How the message is transmitted (must be a credible means of transmission).
■ Message content (what): Information conveyed. (Does the message contain the information needed by the recipient to make a decision?)
■ Recipient (to whom): Who should receive the message. (Who would credibly receive it, and who ultimately needs to receive it to take action?)

All of these variables will influence the action taken (to what effect). Some sample messages follow.

Sample Message #1

From: Police
To: EOC Police Staff

Cab door of trailer truck has been forced open and driver has been removed. Shipping papers indicate hydrochloric acid being transported. Acid flowing into sewers. Attempts to open rear trailer door ongoing.

Sample Message #2

From: Environmental Protection (Field)
To: EOC Environmental Staff

Resident managers of apartments in area request information concerning safety of drinking water, water in swimming pools, and dwellings after evacuees are allowed to return to homes.

Sample Message #3

From: EOC Fire Staff
To: Fire/Rescue Communications

Weather Service reports winds in an east-northeast direction at 10 to 15 mph with gusts of 20 to 25 mph. Forecast for continued rain with possible thunderstorms with strong gusting winds of up to 45 mph during the storm. Current temperature: 82 degrees.

Sample Message #4

From: Chemical Facility Safety Officer
To: Fire Dept. HazMat Team Leader

The supervisor responsible for shutting off chemical valves in a plant area near the fire has not reported in. It is unknown whether the valves have been shut off and whether the person has evacuated.

Sample Message #5

From: Betsy Ames
To: Township Fire PIO

My name is Betsy Ames. I'm a reporter for the Daily Express News. Can you give me a flood situation report for Hamilton, Jordan, Kemper, and the lakes in this area?

```
┌─────────────────────────────────────────────────────────────────────┐
│  EMERGENCY EXERCISE                                                   │
│  <MESSAGE>                                                            │
│            ┌──────────────────┐   ┌──────────┐  ┌──────────────┐     │
│            │  TO:             │   │  METHOD  │  │  FROM:       │     │
│            │                  │   │          │  │              │     │
│            │                  │   │          │  │              │     │
│            └──────────────────┘   └──────────┘  └──────────────┘     │
│                                                                       │
│            ┌──────────────────┐   ┌─────────────────────────────┐    │
│            │  NO:             │   │  TIME:                      │    │
│            │                  │   │                             │    │
│            └──────────────────┘   └─────────────────────────────┘    │
│                                                                       │
│  CONTENT: _____   │
│                                                                       │
│  ACTION TAKEN: _____   │
│                                                                       │
└─────────────────────────────────────────────────────────────────────┘
```

Figure 8.18   Sample message format.

Message Format

Exercise designers often use a standardized form with spaces for the four variables:

1. To
2. From
3. Method
4. Content

A standard message form may also have a space for message number (see Figure 8.18).

Spontaneous Messages

The majority of exercise messages in a functional exercise will be prescripted. And in the course of designing a functional exercise, it is a good idea to try to anticipate things that might go wrong and to provide the controller and simulators with ideas about ways to handle those situations.

In an actual exercise activity, however, the participants don't always respond as you expect them to. Then, it becomes necessary for the controller and simulators to improvise. Usually, the controller will decide on an appropriate response. But if the action is intense, the simulators may also have to make quick decisions and ad lib. If the controller and simulators are familiar with the scenario and objectives, their spontaneous messages can still fulfill the purposes of the exercise.

Composing a Message

In composing a message, begin with an expected action.

- Think about who could send a message and what that person could say to motivate the expected action.
- Think about the four message variables:
  - Who would credibly send the message?
  - How will the message be transmitted?
  - Who will receive the message? If not the decision maker, where would the message be redirected?
  - Does the message provide all of the information needed to make a decision?

Keep it realistic. Practice with the messages. Read them through with someone who is familiar with the organization involved. Does the message motivate the expected action? If it does, then you probably have a successful message.

Pulling It Together: The Master Scenario of Events List

During a functional exercise, a master scenario of events list (MSEL) is often used to monitor the progress of the exercise to keep it on schedule and on track. This chart should provide a picture of the whole exercise. This list should not be shown to participants. Table 8.12 is example of an MSEL worksheet.

Conclusion

In July of 1989 a United Airlines passenger jet crashed in Sioux City, Iowa. The plane burst into flames on the runway after a failed landing attempt. Although 109 passengers died in this terrible tragedy, the lives of 186 passengers were spared because of the response efforts of emergency personnel. Their survival was due mainly to three factors:

- Response of the flight crew before the crash
- Trained rescue units waiting on the ground
- Centralized communications among all response agencies

These factors were present because of training, and the high level of training was no coincidence. Years before the crash, a disaster services center was established. Representatives from forty local agencies met regularly to review emergency procedures and plan realistic exercises.

Table 8.12   Master Scenario of Events List Worksheet Example

Time	Message/Event	Expected Actions
0735	Plane radios tower: losing engine power and altitude.	1. Tower notifies dispatch center. 2. Dispatch alerts police, fire, medical to proceed to airport.
0740–0750	Pilot reports major vibrations/noise; requests runway designation.	1. Tower designates runway; notifies dispatch of runway and potential for mass casualty incident. 2. Dispatch relays runway to police, fire, medical. 3. Dispatch notifies hospitals. 4. Crash fire rescue initiates ICS; notifies dispatch of ICP and staging locations. 5. Dispatch relays ICP and staging locations to police, fire, medical.
0755	Hospital calls dispatch requesting more information.	1. Dispatch obtains potential number of casualties and relays to hospital. 2. Hospital notifies other medical facilities.
0800	Media call dispatch requesting information.	(Etc.)

Two years before the crash, the community conducted a full-scale exercise based on a commercial plane crash. The simulated event revealed several problems, including confusion in communications and inadequate equipment at the scene. An after-exercise plan was developed to address these problems and, as a result, these same problems were not a factor during the real emergency.

Many communities around the nation have had similar experiences that show the value of exercising the EOP. Research has shown that, during a disaster, people will generally respond the way they have been trained. Therefore, it makes sense that exercising will lead to better preparedness and a better overall response.

Chapter 9

Three Types of Exercises

Michael Fagel

Contents

Introduction

Chapter 8 covered the big picture of how an exercise program works. This chapter continues with the importance of exercises and focuses, in detail, on three types of exercises—tabletop, functional, and full-scale. A tabletop exercise simulates an emergency situation in an informal, stress-free environment. The participants usually gather around a table to discuss general problems and procedures in the context of an emergency scenario. The focus of the exercise is on training and familiarization with roles, procedures, or responsibilities.

The Tabletop Exercise

The tabletop exercise—usually simply called a *tabletop*—is largely a discussion guided by a facilitator. Its purpose is to solve problems as a group. There are no simulators and no attempts to arrange elaborate facilities or communications. One or two evaluators may be selected to observe proceedings and progress toward the objectives. The success of a tabletop exercise is determined by feedback from participants and the impact this feedback has on the evaluation and revision of policies, plans, and procedures.

Advantages and Disadvantages

The tabletop exercise is very useful as a training tool, but has both advantages and disadvantages, which are laid out in Table 9.1.

Table 9.1   Advantages and Disadvantages of Tabletop Exercises

Advantages	• Requires only a modest commitment in terms of time, cost, and resources
	• Is an effective method for reviewing plans, procedures, and policies
	• Is a good way to acquaint key personnel with emergency responsibilities, procedures, and one another
Disadvantages	• Lacks realism and thus does not provide a true test of an emergency management system's capabilities
	• Provides only a superficial exercise of plans, procedures, and staff capabilities
	• Does not provide a practical way to demonstrate system overload

How a Tabletop Works

In many respects, a tabletop exercise is like a problem-solving or brainstorming session. Unlike a functional exercise, problems are tackled one at a time and talked through without stress.

Problem Statements and Messages

A tabletop is not tightly structured, so problem statements can be handled in various ways:

■ The facilitator can orally present general problems, which are then discussed one at a time by the group.
■ Problems can be orally addressed to individuals first and then opened to the group.
■ Written detailed events (problems) and related discussion questions can be given to individuals to answer from the perspective of their own organization and role, then discussed as a group.
■ Another approach is to deliver prescripted messages to players. The facilitator presents them, one at a time, to individual participants. The group then discusses the issues raised by the message, using the emergency operations plan (EOP) or other operating plan for guidance. The group determines what, if any, additional information is needed and requests that information. The group may take some action if appropriate.
■ Occasionally, players receiving messages handle them individually, making a decision for the organization that they represent. Players then work together, seeking out information and coordinating decisions with each other.

Some facilitators like to combine approaches, beginning the exercise with general problems directed to key individuals and then passing out messages one at a time to the other players.

Handling problems. It is usually wise to take the time to resolve problems, rather than hurry from one problem or message to the next, even though players sometimes will want to bypass the tough problems.

Facilities and Materials

It is recommended that the emergency operations center (EOC) or other operations center be used for the tabletop exercise for two reasons:

1. It provides the most realistic setting.
2. Needed plans, displays, and maps are available on the premises.

Any conference facility that will comfortably accommodate the expected number of participants in a face-to-face setting will be adequate, however. The number of participants and the scenario will determine the number and arrangement of tables for the exercise. Some facilitators like to arrange small groups around separate tables. Others prefer a U-shaped layout. Provided reference materials should include emergency plans, maps, and other reference materials that would normally be available in the EOC.

Facilitating a Tabletop Exercise

A tabletop exercise provides a relaxed environment of team problem solving. Whereas functional and full-scale exercises are interactive, a tabletop exercise is managed by a facilitator. The facilitator has a number of responsibilities, including:

- Setting the stage (e.g., introducing the narrative)
- Involving everyone—stimulating discussion, and drawing answers and solutions from the group (rather than supplying them)
- Facilitating in-depth problem solving.
- Controlling the pace and flow of the exercise and distributing messages

The facilitator must have good communication skills and be well informed on local plans and organizational responsibilities. Although the facilitator can be thought of as a discussion leader, the role can be much more broad. The following are some guidelines for facilitating a tabletop exercise.

Setting the Stage

The opening remarks and activities influence the whole experience. Players need to know what will happen and to feel comfortable about being there.

Guidelines for Setting the Stage

- Welcome—Begin by sincerely welcoming participants and putting them at ease.
- Briefing—Brief the participants about what will happen. This includes a clear explanation of:
  - Purposes and objectives
  - Ground rules
  - Procedures
- Narrative—Start the exercise by reading (or having someone read) the narrative and introducing the first problem or message.
- Ice breaker—Try breaking the ice by beginning with a general question directed at one or two high-ranking officials or to the group as a whole. Later, other problem statements or messages can be addressed to other individuals or organizations.

Tabletop Participation

It is important that everyone participates and that no one person or department dominates the discussion. Tips for involving all of the participants are summarized in the following list.

Ways to Involve All of the Participants

- Organize the Model and messages so that all organizations must deal with a question or problem.
- Give extra encouragement to those who are a little reticent.
- Avoid the temptation to jump in with the right solutions when players are struggling. This will often hamper the discussion. Instead, try to draw out the answers from the players. They will be more likely to participate if they feel people are listening intently and sympathetically.
- Encourage the behaviors you want from the participants.
  - Give eye contact.
  - Acknowledge comments in a positive manner.

In-Depth Problem Solving

The purpose of tabletop exercises is usually resolving problems or making plans as a group. That means going after real solutions, not superficialities.

Some facilitators make the mistake of trying to move too fast through the scenario, believing that they have to meet all of the objectives and get through all of the messages. That is not a good approach, however, if nothing gets settled.

Remember: If you spend all the time on one big problem, maintain interest among players, and reach consensus, then the tabletop exercise is a success. *Push the players past superficial solutions.* A few carefully chosen, open-ended questions can keep the discussion going to its logical conclusion.

To maintain a high level of interest and keep everyone involved, the facilitator needs to control and sustain the action. There are several ways to do this.

Ways to Control and Sustain Action

- Use multiple event stages—Develop the scenario narrative in event stages. (For example, the initial narrative may involve warning. A later one could deal with search and rescue.) Then, as discussion begins to fade on one issue, introduce the next segment.
- Vary the pace—Add or delete problem statements and messages to alter the speed of the action. Occasionally give two messages at the same time to increase pace and interest.
- Maintain a balance—Maintain a balance between talking a problem to death and moving along so fast that nothing gets settled. Don't hesitate to control the exercise tightly!
- Watch for signs of frustration or conflict—Always remember that the tabletop is basically training, not testing. People may come with fragile egos and little exercise experience. If you see mounting frustration or conflict, stop the exercise. Reach into your experience as a discussion leader to help the players resolve conflicts and feel comfortable.
- Keep it low key—Avoid a bad experience by keeping in mind the low-key nature of the tabletop exercise.

Designing a Tabletop Exercise

The eight-step process discussed in Chapter 8 is used to design a tabletop exercise. The steps are:

1. Assess needs
2. Define the scope
3. Write a purpose statement
4. Define objectives
5. Compose a narrative
6. Write major and detailed events
7. List expected actions
8. Prepare messages

For a tabletop exercise, the process can be somewhat simplified. Because a tabletop exercise is only partially simulated, it requires little scripting. The only roles are the facilitator, the participants (responding in their real-life roles), and one or two recorders. Recorders are used to take minutes and record decisions and usually do not need formal evaluation forms.

Applying the Design Steps

The first four steps are handled as outlined in Chapter 8. The remaining steps can be simplified as follows:

- Narrative—The tabletop narrative is sometimes shorter. It is nearly always given to the players in printed form, although it can be presented via television or radio. When the purpose of the exercise is to discuss general responses, the narrative can be presented in parts, with a discussion of problems after each part.
- Events—The events should be closely related to the objective of the exercise. Most tabletop exercises require only a few major or detailed events, which then can easily be turned into problem statements.
- Expected actions—A list of expected actions is useful for developing problem statements and messages. It is always important to be clear with what you want people to do.
- Messages—A tabletop exercise can succeed with just a few carefully written messages or problem statements. As always, messages should be closely tied to objectives and should be planned to give all participants the opportunity to take part. The messages might relate to a large problem or a smaller problem, depending on the purpose of the exercise. Usually they are directed to a single person or department, although others may be invited to join the discussion.

It is a good idea to write a few more messages than you think you will need. If messages are carefully thought through, they will create a rather lengthy discussion. It's better to have ten or fifteen good messages than twenty or thirty hastily written messages.

The Functional Exercise

The functional exercise simulates an emergency in the most realistic manner possible, short of moving real people and equipment to an actual site. As the name suggests, its goal is to test or evaluate the capability of one or more functions in the context of an emergency event.

It is important not to confuse *functional exercises* with emergency *functions*. All exercises (tabletop, functional, and full-scale) test and evaluate functions contained in the EOP. In this course, *functions* refers to actions or operations required in emergency response or recovery. The thirteen functions recognized by FEMA are:

- Alert Notification (Emergency Response)

- Warning (Public)
- Communications
- Coordination and Control
- Emergency Public Information
- Damage Assessment
- Health and Medical
- Individual/Family Assistance
- Public Safety
- Public Works/Engineering
- Transportation
- Resource Management
- Continuity of Government

The key characteristics of functional exercises, discussed in Chapter 8, are:

- Interactive exercise, designed to challenge the entire emergency management system; can test the same functions and responses as in a full-scale exercise without high costs or safety risks; usually takes place in an EOC or other operating center.
- Involves controller(s), players, simulators, and evaluators.
- The atmosphere is stressful and tense because of real-time action and the realism of the problems.
- Exercise is lengthy and complex; requires careful scripting, careful planning, and attention to detail.
- Geared for policy, coordination, and operations personnel (the players).
- Players practice their response to an emergency by responding in a realistic way to carefully planned and sequenced messages given to them by simulators.
- Messages reflect a series of ongoing events and problems.
- All decisions and actions by players occur in real time and generate real responses and consequences from other players. Guiding principle: imitate reality.

Best Uses for the Functional Exercise

The functional exercise makes it possible to test the same functions and responses that would be tested in the full-scale exercise, without the high costs or safety risks. The functional exercise is well suited to assess the:

- Direction and control of emergency management
- Adequacy of plans, policies, procedures, and roles of individual or multiple functions
- Individual and system performance
- Decision-making process
- Communication and information sharing among organizations

- Allocation of resources and personnel
- Overall adequacy of resources to meet the emergency situation

Table 9.2 compares tabletop and functional exercises.

Participant Roles

As already noted, the functional exercise involves players, simulators, a controller, and evaluators. In a small jurisdiction, one or two people may serve as controller, simulator, and evaluator. In a larger jurisdiction, these roles may be filled by many people.

Players

The players in a functional exercise are people who hold key decision-making or coordinating positions and would normally function in the operations center.

By operations center, we mean the central location that is designated in a real emergency for policy decisions, coordination, control, and overall planning. For a governmental jurisdiction, it would be the EOC; for a volunteer agency or private sector entity, it would be the central location from which key decision makers operate in an emergency situation.

Decision makers. Key decision makers would normally include leaders in government and key responding organizations: the mayor or other chief executive; chiefs and coordinators of emergency services such as fire, police, EMS, public information officer (PIO); and so on. In a nongovernmental organization, the chief executive officer and other organizational leaders would participate.

Coordination and operations. Serving in the coordination and operations groups are people from various departments who work with policy makers. In large exercises, a separate operations group carries out directives. In small exercises, the coordination and operations roles may be taken on by the policy makers. The best guide in selecting who should participate in an exercise is the emergency plan.

Duties. The only job of the players is to respond as they would in a real emergency to the messages that they receive during the exercise. All of the decisions and actions of the players take place in real time and generate real responses and consequences from other players.

Simulators

To create a real-life environment, simulators portray the organizations that would normally interact with the players in the operations center. They do this by delivering messages/descriptions of events or problems that require players to act.

Some messages are scripted in advance; others are spontaneous responses to player decisions. They are input into the exercise by means of radio or telephone, or by written notes simulating radio and telephone transmissions.

Table 9.2   Tabletop versus Functional Exercises

	Tabletop	*Functional*
Degree of realism	Lacks realism	As realistic as possible without deploying resources
Format/structure	Group discussion, based on narrative and problem statements/messages	Interactive; simulators deliver "problem" messages, players respond in real time
Atmosphere	Low-key, relaxed	Tense, stressful
Who takes part	Facilitator, participants (decision-making level); may use recorders	Controller, players (policy, coordination, and operations personnel), simulators, evaluators
Who leads	Facilitator	Controller
Where held	EOC, other operations center, or conference room	EOC or other operations center
Equipment deployed	No	No
Test coordination	Yes, on a discussion level	Yes
Test adequacy of resources	No	Yes
Test decision-making process	Yes	Yes
Relative complexity/cost	Small group, simple format, modest cost	Large scale, complex format, moderate cost to design and implement (higher than tabletop, lower than full-scale)
Formal evaluation	No (self-assessment by participants)	Yes

Duties: Simulators are responsible for all actions taken by organizations or individuals outside the EOC. They:

- Send the players prescripted messages representing private citizens, agencies, or other organizations, according to scheduled times in the sequence of events
- Simulate all actions taken by an agency or other organization
- Ad lib spontaneous messages as needed. Examples of times when a simulator may need to respond spontaneously include:
  - When a member of the operations center issues a directive that results in events not anticipated in the scenario
  - When a player asks for more information
  - When a player's decision is not logically linked to the next event in the scenario. Inform the controller of any deviations from the scenario, or special problems.

When simulators are given directives, they are required to follow through and implement the directives in a professional manner.

Selection: Simulators must be able to ad lib intelligently in the situations just described, so it is important that they be familiar with the organization(s) that they are simulating and with the sequence of events and messages. It is useful, therefore, to draw simulators from the organizations that they will portray or from the design team.

Numbers: It is difficult to give a rule of thumb concerning specific numbers of simulators needed for an exercise. The number of simulators will vary according to the:

- Number of players
- Length of the exercise
- Knowledge and training of the simulators
- Communication channels available

For the best results, try to have at least one simulator per organization represented in the operations center, with extras to play the part of citizens or other private organizations.

Organizing. It is a good idea to group simulators according to function, to simplify the exercise and reduce the number of simulators needed. One approach is to organize them into three groups:

1. Government agencies not participating in the exercise
2. Participating organizations: Field units of organizations participating in the exercise (police, fire, public works, etc.) and private medical and support organizations
3. Other private facilities and individuals: Citizens and non-government organizations

Table 9.3 illustrates how this approach could be used for a community.

Table 9.3 The Benefits of a Simulation for a Community

Nonparticipating Government Entities	Participating Organizations	Other Private Facilities/Individuals
One or two persons simulating: • Federal regulators • State or area EOC • County EOC • Other city EOC • State/federal officers • Care and shelter • Resources and support	One person per organization simulating: • City departments and agencies • County departments • Medical/health services • Volunteer organizations	One or two persons simulating: • Industries • Commercial businesses • Media • Private citizens

Controller

The controller supervises the simulation or overall conduct of the exercise, making certain that it proceeds as planned and that objectives are reached. The controller must be able to view the exercise as a whole and to think quickly on his or her feet. Players often make unanticipated decisions, and the controller must be able to respond to these.

Duties. The main duties of the controller are to:

- Ensure that the simulators and evaluators are properly trained before the exercise
- Orient the participants to the exercise and present the narrative;
- Monitor the sequence of events and supervise the input of messages, using the master scenario of events list as a guide
- Adjust the pace of the exercise when needed—inserting more messages when it drags and discarding messages when the pace is too frantic
- Make decisions in the event of unanticipated actions or resource requirements
- Maintain order and professionalism throughout the exercise

Selection. Controllers can usually be drawn from the exercise design team. Because the team members are already familiar with the exercise, they are well suited to the task of keeping the exercise moving toward the anticipated conclusion.

Preparation. To prepare properly for the event, the controller should have the following items available:

- List of objectives
- Master scenario of events list
- Messages

- List of players
- List of resources available to the jurisdiction or organization

It is usually helpful to hold a briefing before the exercise to orient the staff members. At the briefing, the controller should train the simulators, ensuring that they are familiar with the scenario, objectives, resources, and messages that they will be responsible for delivering. The evaluation team leader should provide similar training to the evaluators, including exercise objectives, evaluator duties, and schedule.

Evaluators

The evaluators observe the actions and decisions of the players to report later what went well and what did not. To do this, evaluators need to be familiar with the objectives, the exercise scenario, and the jurisdiction or organization that is undertaking the exercise.

Duties. Key duties of the evaluators include the following:

- Observing exercise progress and recording observations (usually on provided evaluation forms), taking care to remain unobtrusive in the process
- Noting how well the exercise is fulfilling objectives and trying to identify problems if objectives are not met
- Evaluating the actions of the players, not the players themselves; documenting both positive and negative observations.
- Informing the controller during the exercise of any problems
- Preparing brief, written comments that can be included in the final evaluation and recommendation report that will be prepared by the emergency manager or other responsible party

How a Functional Exercise Works

A brief review of how a functional exercise works will be provided; however, you will gain a better understanding of a functional exercise if you look for an opportunity to observe one—or better yet, participate in one.

The Beginning

When a functional exercise begins will depend on its objectives. If testing the notification function is one of the objectives, then a "no-notice" exercise is useful. In this case, participants are given only the approximate timeframe scheduled for the exercise—(anywhere from one day to several weeks). The exact time when it begins will be a surprise, allowing the exercise evaluators to observe how effectively notification and assembly at the command point take place. In exercises where notification is not an objective, the exercise time is usually announced in advance.

Briefing

Exercise participants may arrive at a functional exercise with only a vague notion of what is to take place. The exercise is much more likely to be successful, however, if the participants receive a briefing that covers:

- Overview of the objectives
- How the exercise will be carried out
- The time period to be simulated
- Ground rules and procedures

Keep the Briefing Short

Avoid anything that distracts from the atmosphere of a real emergency. (For example, include a written announcement in the exercise materials to cover any administrative details such as restrooms and break times.)

Narrative

The exercise formally begins with the presentation of a narrative. It can be read aloud, dramatized, or presented via television, computer, radio, or slides.

Message Delivery and Response

The action begins as simulators and players interact with one another:

- Simulators communicate messages to players, and players respond as they would in a real emergency.
- Players make requests of simulators, and simulators react convincingly.

This ongoing exchange takes place according to the carefully sequenced scenario of events that governs what takes place, when each event occurs, and the messages used to inform the players.

Messages can arrive on paper, by telephone, by radio, or in person. Using telephones, when possible, increases the feeling of a real emergency, but whispered messages or written notes can also work well.

The success of the exercise depends on the extent to which the participants are able to carry out their functions as if they were in a real emergency. Exercise participants should be encouraged to think of each message as an actual event.

Encouraging Spontaneity

The players should be able to decide among the full range of responses that are normally available to them during an emergency. Their ability to make decisions, communicate, or otherwise carry out their responsibilities should not be constrained by the exercise situation.

To allow the participants spontaneity, exercise controllers and simulators must be well trained and prepared to handle the unexpected. While this requirement provides a better exercise for participants, it does place a burden on controllers and simulators who must be ready to go with the flow to some degree when the situation calls for it.

Controlling the Action

While simulators and players are transmitting messages and responding to them, the controller carefully monitors the interaction and progress.

Dealing with spontaneous decisions. The controller should be made aware of significant spontaneous decisions and make adjustments in the scenario where necessary.

Adjusting the pace. The controller can control the pace of the exercise by adjusting the message flow—slowing things down when the pace is too frantic or speeding it up when the exercise drags. The controller can also even out the pace among participants. Remember, one inactive organization can distract others and lower the intensity of the exercise. Avoid boredom by ensuring a smooth flow of messages.

Strategies for Adjusting the Pace of Messages

Slow the pace by:

- Rescheduling events to allow more reaction time. Have the simulators wait before sending messages.
- Discarding messages that are relatively unimportant or do not greatly impact other decisions. Throw away messages that don't contribute to the objectives.

Increase the pace and fill gaps by:

- Speeding up the delivery pace (varying from the planned schedule).
- Determining what is causing gaps and being ready to add or alter messages spontaneously when needed. Look at organizations with gaps to see if they have been unintentionally ignored. If so, add messages. (It may be, however, that the organization simply has little to do during a particular period.)
- Keeping a supply of optional messages on hand that can be added when needed.
- Adding side events—routine actions that a department would have to continue throughout an emergency. (For example, insert a routine traffic accident to put stress on police and fire departments. Report an unrelated heart attack to challenge medical personnel.)

Table 9.4   Sample Message Flow Chart

Check the times when messages are scheduled for delivery to each organization.						
	Participating Agency/Organization					
	Fire	*EMS*	*Public Works*	*EOC*	*Facility CEO*	*School*
Exercise Start						
1000	✓	✓				
1003			✓			
1006		✓		✓		
1009	✓			✓		
1012			✓	✓		
1015		✓			✓	✓
Etc.						✓

- Adding secondary emergencies—events that develop out of the main flow of exercise events. (For example, insert utility outages, water main breaks, gas leaks, media calls, and similar events to keep players involved between their own major events.)
- Adding special planning requirements that would cause an inactive group to engage in a short-term preparedness activity. (For example, have hospitals test emergency generators.)
- Adding misdirected messages—messages given to the wrong agency. Such messages can be used to gauge the agency's clarity of role definition and to test whether they forward the message properly.

Relieve overloads on particular organizations by:

- Reassigning—Verify that all messages are assigned to the right organizations. Then reassign any messages that could be used by another organization.
- Thinning—Divide the overloaded messages into two piles: (1) essential to the flow of the exercise and (2) nice to have. Then get rid of some from the latter group.

Maintain an even message flow by maintaining a chart similar to Table 9.4.

Skipping Time

Functional exercises can depict events and situations that would actually occur over an extended time period (one or two weeks or more). To include multiple phases of the emergency (preparation, response, recovery, mitigation) in a two-day exercise, it would be necessary to stop the exercise periodically and advance the time by a number of hours or days. These skip-time transitions should be kept to the minimum necessary to cover the scope of the exercise. They can usually be planned to coincide with a natural break point.

Who handles the time skips? The controller is responsible for managing skip-time transitions and preparing transition updates to be presented to the participants before resuming the exercise.

Simulators are responsible for updating simulation displays to reflect the results of the previous events and participant actions. Actions that would have been undertaken during the transition period will be indicated as accomplished on the transition date.

Figure 9.1 illustrates a skip-time schedule for a functional exercise.

Location

Exercise where you operate. To the extent possible, the functional exercise should take place in the same facility and in the same operational configuration that would occur in a real emergency—usually in the EOC.

A frequent objection to exercising at the EOC is that there are not enough chairs, phones, or restrooms. If that is the case, it is wise to find out during an exercise. If you can't practice there, don't expect to be able to conduct an emergency response there.

Room Arrangement

Various room arrangements can work for a functional exercise, depending on the size of the exercise. These are the basic requirements:

- Space for players—usually a table with plenty of workspace
- Areas set aside for simulators
- Room for evaluators to observe
- A place from which the controller can operate

In very small exercises, a single room can work. Figure 9.2 shows a simple layout for a small functional exercise. Figure 9.3 shows a layout that would be more appropriate for complex exercises. Two rooms are shown, the simulation room and the operations center, where the players are located.

If more than one or two departments or functions are being exercised, a simulation room is highly recommended. This room should comfortably house all of the simulators so that they can send, receive, and track messages and other communications with the players. It should be equipped with telephones or radios if they are to be used in the exercise. If message traffic is to be sent by hand, the situation room must be near the players.

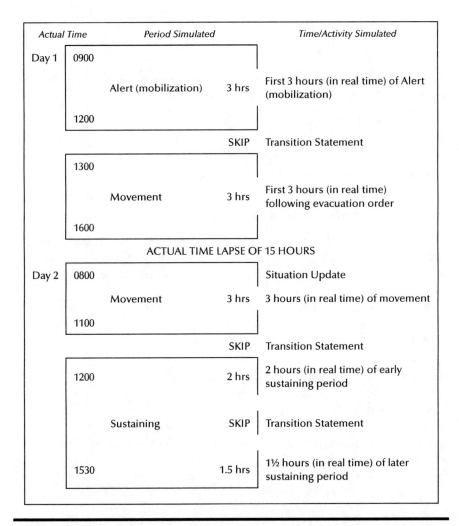

Figure 9.1   Sample skip-time schedule for a two-day functional exercise.

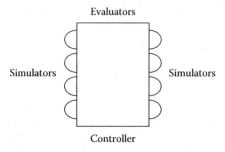

Figure 9.2   Sample arrangement for a small functional exercise.

Any layout should be adapted to the particular exercise and your local physical facilities. Whatever the layout, participant work spaces should be predesignated and working supplies made available. (In the previous diagram, notice the workspace assignments of the simulators and players.)

Communications Equipment

Communications equipment is useful when full simulation is the goal. However, often it is wise to use both electronic equipment and written messages. When working with a compressed time period, it is easy for problems to arise concerning:

■ Development of telephone banks for the simulators
■ Telephone overload for the players
■ Equipment breakdown

For these reasons, some emergency managers leave extensive communications equipment for a drill. In any case, the use of electronic communications should be carefully—and perhaps selectively—planned.

Equipment installation. When telephones will be the primary means of communication during an exercise, it may be possible to use existing phones. Or, it may be necessary to install special lines and extensions to provide the necessary communication links. In some facilities, where a central switching system is used, an operator may handle all calls.

Suggestions for Successful Communication Links

■ Prepare a special exercise directory of telephone numbers
■ Include communications procedures in the directory
■ If you don't have telephones, use a variety of other formats, such as:
■ Written messages
■ Simulated calls (sender whispers message in receiver's ear)

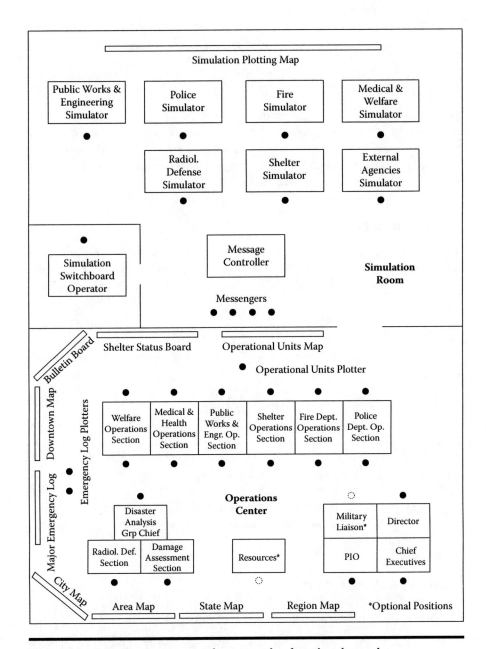

Figure 9.3   Sample arrangement for a complex functional exercise.

- Hand signals (player who wants to call a simulator raises a hand to bring the simulator over)
- Simulated speaker phone or radio (simulator speaks loudly to the players)
- If you use written messages, provide standardized message forms

Displays and Materials

Displays and materials—maps, charts, message forms, lists, and so forth—are important in a functional exercise. These materials are used to provide details for the scenario and keep track of activities. Generally speaking, it's best to use what you use every day. An exercise is no time to get new maps and message forms. Test the materials that you are currently using.

Design Process

The full eight-step design process outlined in Chapter 8 is used to develop a functional exercise. Although a simplified version of the process can be used to develop a tabletop exercise, a functional exercise requires careful attention to every step.

Exercise Materials

The success of a functional exercise rests on a carefully scripted scenario package that includes:

- A convincing narrative
- Major and minor events that grow out of the narrative and are carefully chosen to support the objectives
- Arrangement of the events in a realistic and convincing sequence from the beginning to the end of the exercise
- Expected actions that are tied closely to the objectives
- A great number of specific messages (perhaps one hundred or more in a larger exercise) that are so well conceived that players will respond with the expected actions

Expect the Unexpected

No matter how good you are at writing convincing messages, sometimes players will respond in unexpected ways. Although you should try to limit the unexpected as much as possible, occasionally a spontaneous reaction is better than the response prescribed in the emergency plan. To that end, you should have a master scenario of events list that includes all of the messages/events, delivery times, and expected actions.

When you have completed the scenario package, you will use the developed materials to create materials for the exercise participants, including the exercise plan, control plan, evaluation plan, and player handbook.

Figure 9.4 will help you prepare for a functional exercise.

The Full-Scale Exercise

The full-scale exercise is the culmination of the exercise process—the exercise that will depict, most accurately, a real emergency. A full-scale exercise is as close to the real thing as possible. It is a lengthy exercise that takes place on location, using—as far as possible—the equipment and personnel that would be called upon in a real event.

In a sense, a full-scale exercise combines the interactivity of the functional exercise with a field element. It differs from a drill in that a drill focuses on a single operation and exercises only one organization. Eventually, every emergency response organization must hold a full-scale exercise because it is necessary at some point to test capabilities in an environment as near to the real one as possible.

However, there is more to a full-scale exercise than just practice in the field. As we discussed earlier, various regulatory agencies have requirements for full-scale exercises that must be satisfied. To receive FEMA credit, for example, a full-scale exercise must fulfill three requirements:

1. It must exercise most functions.
2. It must coordinate the efforts of several agencies.
3. To achieve full coordination, the EOC must be activated.

Key Characteristics of the Full-Scale Exercise

The key characteristics of a full-scale exercise were discussed in Chapter 8. For review, here is a brief summary of the main points.

■ Interactive exercise, designed to challenge the entire emergency management system in a highly realistic and stressful environment.
■ Tests and evaluates most functions of the emergency management plan or operational plan.
■ Takes place in an EOC or other operating center and at field sites.
■ Achieves realism through:
  – On-scene actions and decisions
  – Simulated "victims"
  – Search and rescue requirements
  – Communication devices
  – Equipment deployment
  – Actual resource and personnel allocation

Facilities and Equipment
☐ Sufficient work space for simulators and players
☐ Simulation room (if needed) near player room
☐ Message center
☐ Control center
☐ Observer space
☐ Clear work surfaces
☐ Communication equipment (telephones, switchboard)
☐ Parking
☐ Adequate ventilation and lighting

Displays and Materials
☐ Displays easily visible or accessible
☐ Maps (regional, state, local, area, downtown, operational units)
☐ Major events log
☐ Bulletin board
☐ Status boards
☐ Simulation plotting board
☐ Easels, chart paper
☐ Message forms
☐ Pencils/Paper
☐ Name cards

Beginning
☐ "No-notice" or scheduled (according to objectives)

Briefing (short)
☐ Objectives
☐ Process
☐ Time period portrayed
☐ Ground rules and procedures

Narrative
☐ Verbal, print, TV, computer, slides, or dramatization
☐ Time-skips if needed

Messages
☐ Large number (depends on scope)
☐ Prescripted
☐ Optional prescripted for adjusting flow

Message Delivery
☐ Written
☐ Phone
☐ Other (oral, speaker phone/radio, hand signals)
☐ Simulators prepared for spontaneous message development
☐ Standardized forms for written messages

Strategies for Adjusting Pace
☐ Rescheduling
☐ Adding/deleting messages
☐ Misdirecting messages
☐ Reassigning messages

Figure 9.4 Functional exercise design checklist: special considerations.

- Involves controller(s), players, simulators (different from simulators in a functional exercise), and evaluators.
- Players represent all levels of personnel, including response personnel.
- Messages may be visual (e.g., staged scenes, made-up victims, props) and scripted.
- All decisions and actions by players occur in real time and generate real responses and consequences from other players.
- Requires significant investment of time, effort, and resources (1 to 1½ years to develop a complete exercise package). Attention to detail is crucial.

The Purpose of a Full-Scale Exercise

There are numerous reasons for conducting a full-scale exercise. A full-scale exercise:

- Enables a jurisdiction or emergency management system to evaluate its ability to perform many functions at once.
- Is useful to test total coordination, not only among policy and coordination officials, but among field forces. At the same time, it can test interorganizational coordination.
- Can pinpoint resource and personnel capabilities and reveal shortfalls.
- Greatly expands the scope and visibility of the exercise program.
- If well-planned, can attract public attention and raise credibility. (To be successful, however, it must be the culmination of a comprehensive and progressive exercise program that has been developed as the jurisdictional capacity has grown.)

Requirements to Run a Full-Scale Exercise

Some people wrongly believe that when a full-scale exercise is started, it can run on its own steam. In fact, a full-scale exercise requires a substantial commitment of time, money, personnel, and expertise, and should not be undertaken without the necessary preparation. These are the most important requirements:

- Substantial experience with preparatory exercises of various kinds—drills, tabletops, and functional exercises
- Total commitment of all emergency service organizations
- Support from the chief elected and appointed officials
- Adequate physical facilities, including space for the EOC and field command posts
- Adequate communication facilities (e.g., radios and telephones)
- Plans in place to handle costs (both evident and hidden), labor, time commitment, and so forth
- Carefully thought-out and planned site, and logistics

Table 9.5   Functional versus Full-Scale Exercises

	Functional	*Full-Scale*
Degree of realism	As realistic as possible without deploying resources	As realistic as possible; resources deployed
Format/structure	Interactive; simulators deliver "problem messages," players respond in real time	Interactive; simulators play roles at the scene, players respond
Atmosphere	Tense, stressful	Highly tense, stressful
Who takes part	Controller, players (policy, coordination, operations), simulators, evaluators	Controller(s), players (all levels), simulators, evaluators
Who leads	Controller	Controller(s)
Where held	EOC or other operations center	EOC and field site(s)
Equipment deployed	No	Yes
Test coordination	Yes	Yes
Test adequacy of resources	Yes	Yes
Test decision-making process	Yes	Yes
Relative complexity/cost	Large scale, complex format, moderate cost	Very large scale, highly complex, high cost
Formal evaluation	Yes	Yes

Table 9.5 briefly compares functional and full-scale exercises.

Full-Scale Exercise Roles

Full-scale exercises involve one or more controllers, the participants, simulator, evaluators, and a safety officer.

Controllers

One or more controllers manage the exercise. In some cases, where there are multiple sites or jurisdictions, there may be more than one controller. In this case, all of the controllers cooperate under the direction of a chief controller. The controller (or chief controller) is responsible for ensuring that the exercise starts on schedule. The controller also designates an exercise control point from which all communications should be monitored.

Participants

A full-scale exercise involves all levels of personnel, including:

■ Policy makers—Those who are responsible for making broad policy decisions. They might include the chief executive and his or her staff, the public information officer, the emergency manager, key department heads, and other elected officials.
■ Coordination personnel—People from various departments who coordinate decisions of policy makers and make plans for resources.
■ Operations personnel—Those who carry out the directives. Sometimes coordination and operations are the same.
■ Field personnel—Fire, police, EMS, search and rescue, volunteer groups, representatives of private enterprises who participate in the response, and many others.

Simulators

Simulators in a full-scale exercise are different from those in a functional exercise. In a full-scale exercise, simulators are volunteers who pretend to be victims of the emergency event. For realism, they may act injured, unconscious, hysterical, or dead—whatever the scene calls for.

Evaluators

Evaluators observe the action and keep a log of all significant events. This is important because so many of the actions will not be prescribed but spontaneous responses to other actions.

Safety Officer

There are so many potential safety issues in a full-scale exercise that a safety officer should be designated. This person's primary responsibility is to analyze the entire exercise from a safety perspective.

How the Full-Scale Exercise Works

The full-scale exercise begins in a fashion similar to the functional exercise; whether it is announced or "no notice" depends in part on the objectives. The exercise designer will decide how and when the exercise is to begin. The trigger may simply be a call from dispatch, a radio broadcast, or a telephone call from a private citizen. The beginning for each participant should be as realistic as possible (that is, personnel should receive notification through normal channels).

Personnel from the emergency services that are taking part in the field component must then proceed to the assigned location, where a "visual narrative" is displayed before them in the form of a mock emergency to which they will respond.

Key decision makers—those who would normally operate out of the EOC or command center during an emergency—proceed to the EOC to fulfill their roles. Command posts are set up as required by the event.

Action

Actions in a full-scale exercise occur in the EOC, at one or more field sites, and at the related command posts. Actions taking place at the event site and command posts serve as input to the simulation taking place at the EOC. Although medical personnel, hospitals, EMS, fire services, and other localized emergency operations do not usually require centralized command from the EOC, they do require coordination with officials at the command posts.

Sustaining Action

Action is sustained by various means, including:

- Prescripted messages input by the controller(s)
- Messages and actions from the field that require action at the EOC
- Spontaneous responses to the various messages and actions

Command Post Messages

A field command post can be used as part of the message input into the EOC. Either the command post can be written into the scenario and have a set of prescripted messages to be transmitted by radio, or the command post controller can monitor the sequence of events and transmit spontaneous messages.

Exercise Locations

Field Sites

The main event site will depend on the exercise scenario and objectives. For example, if the central event involves a plane crash, the exercise might take place at an airport. A simulated terrorist attack could be staged in a public facility such as a convention center or shopping mall. A hurricane or flood exercise might be dispersed over many locations.

Most events will involve additional locations, such as secondary event sites, hospitals, mortuaries, shelters, and other support locations, and command posts will be established near the emergency sites. In fact, one of the reasons that a full-scale exercise is so complicated is that activity is taking place in various locations, and all of the sites must be coordinated.

Emergency Operations Center

Activation of the EOC or other operations center is central to a full-scale exercise (just as it is in a functional exercise). The purpose of the EOC is to provide a policy and coordination facility for the CEO and staff to respond effectively to an emergency.

In essence, the EOC is the voice of government during an emergency. Emergencies place strains on government—the demand for service escalates, while the ability to deliver diminishes. Gathering information, making decisions, and directing necessary actions require close coordination among key officials. This coordination is best obtained if officials and support staff are in a centralized location with direct lines of communications.

The central location makes it possible to accomplish a number of tasks that would be impossible (or very difficult) separately. For example:

- Information can be gathered, verified, and recorded in one spot.
- Officials can deploy resources in a timely and intelligent manner.
- Direction and control can be managed efficiently.
- Officials can coordinate actions and decisions.

It's easier to set meaningful priorities when the key actors collaborate.

The size and makeup of the EOC differ according to the size of the jurisdiction or response system. The EOC may take up an entire floor of a building or it may be located in a small room.

Designing the Full-Scale Exercise

The design of a full-scale exercise can be quite difficult, often requiring the expertise of several response organizations. When developing your first full-scale exercise, it is usually advisable to start small and build to more complex exercises. Many of the potential difficulties relate to logistical problems, but others rest with the design.

The entire eight-step design process is used to design full-scale exercises, although the process is applied somewhat differently when designing tabletop and functional exercises. The differences arise from the fact that tabletop and functional exercises rely on *words* to obtain realism, whereas a full-scale exercise—like a real emergency—gains its reality from *things*. There is a shift from a written scenario to visual reality represented by a real site, real people (some of them simulating victims), and real equipment.

Let's take a closer look at how each of the design steps is applied to designing a full-scale exercise.

The First Four Steps

As with any exercise, the first four design steps are:

1. Assess needs
2. Define the scope
3. Write a statement of purpose
4. Define objectives

For a full-scale exercise, these steps require deeper analysis and greater attention to detail. So much rests on the outcomes of these steps that you must be sure that you have created a clear picture of what is to be achieved through the exercise. If any of these areas is left at all vague, the problems will be greatly magnified later.

Step 5: The Narrative

The narrative is handled differently in a full-scale exercise. Because a lengthy oral description is not needed to set the scene, the narrative is shorter.

Step 6: Major and Detailed Events

Major and detailed events still exist in a full-scale exercise, but many of them exist as actual occurrences rather than as oral descriptions. For example, an earthquake scenario may have to rely on words to simulate some events. Other earthquake events can be simulated with fallen logs, bricks strewn around a building, dummies resting under beams, people acting injured or frightened, and other props. Even when events are presented visually, they cannot be random and haphazard. Each event must be carefully planned and staged to support objectives and generate the expected actions.

Step 7: Expected Actions

As with any exercise, expected actions must be specifically identified, based on the exercise objectives. A detailed list of expected actions is an important foundation for the exercise evaluation.

Step 8: Messages

There are two kinds of messages in a full-scale exercise: visual and prescripted. Much of the action grows out of the initial message and scene setup. The scene that is set up contains a number of "visual" messages to which participants need to respond. It may also be necessary to have some prescripted messages to move the action along.

For every event, try to anticipate all possible reactions. But it is important to be flexible enough to adapt to player actions and decisions that you hadn't expected. Sometimes an unanticipated response will be an improvement over the expected action.

If your scenes (events) are well planned, the exercise will go in the general direction that you planned, even if a few of the decisions are off course.

Watch the details! Make sure that the scenes that you create are good enough to get the expected action. For example, if you have victims who don't know anything about medicine, either coach them in advance about their symptoms or tag them with symptoms and vital signs and apply makeup to simulate injuries. At the end of the exercise, you don't want players to say, "Well, I didn't know what you were getting at."

Special Considerations

A full-scale exercise represents a huge logistical challenge, and it's easy to overlook details. One way to promote clear and creative thinking is to "walk the site"— either physically or mentally. By doing so, you may be able to identify potential problems and do more realistic planning. In mentally evaluating the scene, you should consider:

- Site selection
- Scene management
- Personnel and resources
- Response capability
- Safety and legal liability
- Emergency call-off
- The media

Site Selection

Selecting the right site for the exercise should be one of the first decisions you make. Because the exercise requires the mobilization of both personnel and resources, space and realism are key.

Site Selection Criteria

- Credibility—Is the type of emergency one that has a real possibility of occurring? (A credible emergency can elicit greater cooperation and participation.)

- Realism—Is the site as realistic as possible without interfering with normal traffic or safety?
- Examples—In simulating an overturned tanker truck on a main stretch of freeway, you can't tie up rush-hour traffic for hours. You'll need to find a similar location to stage the crash. For a plane crash, try a secondary airport instead of a national airport.
- Adequacy of space—Is it large enough to accommodate the number of victims, responders, and observers? Is there space for responders' and observers' vehicles?

Scene Management

Management of the scene refers to a number of issues, including:

- Logistics at the scene
- Creation of a believable emergency scene
- Number of victims
- Management of props and materials
- Number of controllers

Figure 9.5 includes questions that you should consider related to scene management.

Logistics
- ☐ Where will players be set up?
- ☐ If there is a mobile EOC, where will they park?

Believability
- ☐ How will you simulate the emergency? (For example, if you will simulate fire, what will you use for smoke? How will you simulate a spilled chemical, broken glass, flood damage, etc.?)
- ☐ Who will serve as victims? (For greater realism, select victims from different age groups, with different body types and physical characteristics.)
- ☐ How will you ensure that the victims realistically portray their roles?

Number of Victims
- ☐ How many victims does the type of emergency call for?
- ☐ What are the capability and capacity of hospitals and other resources to handle victims?
- ☐ What does the history of past events indicate about types and numbers of injuries?

Props and Materials
- ☐ What kinds of props and materials will be needed to simulate injuries, damage, and other emergency effects (e.g., victims' makeup, dummies, construction materials)?

Controllers
- ☐ How many controllers will be needed to manage the exercise sites? (In a multiple-site exercise, every site will require a controller.)

Figure 9.5   Scene management checklist.

Personnel and Resources

The scenario will help determine how many people (participants and volunteers) will be involved, how many and what kinds of equipment will be needed, and the potential costs. Consider the following factors in planning for personnel and resources:

- How many participants? (Sometimes it is necessary to scale down the exercise to a half day or less to increase participation.)
- How many volunteers?
  - Scene setup
  - Victims
  - Members of the public
- What kinds of equipment will be used?
- How many pieces of each type of equipment?
- How much fuel for vehicles and equipment?
- What kinds of materials and supplies will be needed, and how will they be obtained?
- Expenses:
  - Overtime wages
  - Vehicle and equipment fuel
  - Materials and supplies

Equipment

Keep your scenario reasonable in terms of equipment. True, you need to determine whether you have the resources to meet normal emergency demands. But don't get carried away. Hold your people to the use of actual equipment. Don't let them simulate the use of equipment that doesn't exist.

Response Capability

In planning both personnel and resources, take into consideration how the exercise might deplete the actual response capability of the organizations that are involved. It is unwise to compromise the ability to respond to real emergencies during an exercise. Here are some suggestions:

- Be sure that there are enough personnel and resources to continue their responsibilities if a real emergency occurs. (In some cases, a call-off procedure will solve part of the problem.)
- Consider using second-shift personnel or mutual aid from other jurisdictions or organizations.
- Consider using volunteers as a smaller response shift.

Safety

Total operational safety is an absolute must in a full-scale exercise. Awareness of safety issues must be carried throughout the planning and conduct of the exercise to ensure that safety problems are noted and eliminated. A safety officer should be designated whose primary responsibility is to analyze the entire exercise from a safety perspective.

Suggested Safety Measures

- Include safety as one of the activities in exercise development.
- Assign each exercise team member the responsibility of examining the exercise for safety within his or her discipline.
- Identify all possible safety hazards and resolve each one.
- Address safety as part of the preexercise briefing.
- Include safety factors in simulator and evaluator information packets.
- Examine each field location before the exercise to guarantee that safety precautions have been taken.
- Ensure that the safety officer has the authority to terminate an activity—or even the entire exercise—if a safety problem arises.
- Provide for call-off procedures in the event of an actual emergency.

Emergency Call-Off

Be sure to keep enough personnel in reserve to handle routine problems. Be aware, though, that a real emergency may occur during any exercise—especially a lengthy one. In some instances, it may be necessary to stop the exercise to handle the real emergency. Every exercise should have a planned call-off procedure that will result in the prompt return of personnel and equipment to full-duty status. The procedure should include a code word or phrase that the controller or safety officer can use to indicate:

- The exercise has been terminated.
- Personnel should report to their regular-duty positions.
- All radio traffic will return to normal.

The procedures should be tested.

Legal Liability

Legal questions of liability, including liability for injuries during exercise conduct, must be researched by your local attorney.

The Media

A full-scale exercise of any magnitude will draw media attention whether you want it or not. If the exercise is well designed, favorable media reports are more likely. So, include the media in your plans. They can be very helpful in gaining support for the program, and their presence will increase realism.

Also make allowances for observers and public information people. Decide in advance where you will locate them, and give them an opportunity to observe.

Homeland Security Exercise and Evaluation Program (HSEEP) evaluations and guidelines will help you as your jurisdiction prepares to conduct a full-scale exercise.

Exercise Evaluation

Evaluation is the process of observing and recording exercise activities, comparing the performance of the participants against the objectives, and identifying strengths and weaknesses. Evaluation is a complex topic, and this chapter provides only a general overview. We will briefly discuss the importance of evaluation and its relationship to exercise development, evaluation team structure and duties, key aspects of evaluation methodology, and evaluation tasks that happen after the exercise is finished.

Integrating Evaluation throughout Development

Exercise development is a continual process that begins long before the exercise and continues after the exercise is over. It continues until needed changes have been made and suggestions have been incorporated into the next exercise.

Evaluation is not something that is done when all of the action is over. It begins when exercise design begins, when objectives are developed. In Table 9.6 notice the kinds of evaluation activities that occur in the three exercise phases. In the remainder of this chapter, we will take a brief look at each of the major evaluation activities listed in Table 9.6.

Why Evaluate the Exercise?

For an emergency management system to be effective, it is critical that the personnel, plans, procedures, facilities, and equipment be exercised and tested on a regular basis. Yet no amount of exercising will be constructive unless each exercise is followed by a structured evaluation that enables the emergency management organization to identify successes and shortfalls.

Table 9.6   Exercise Evaluation Categories

Task Categories	Preexercise Phase	Exercise Phase	Postexercise Phase
Design	• Review plan • Assess capability • Address costs and liabilities • Gain support/ issue exercise directive • Organize design team • Draw up a schedule • Design exercise (eight design steps, including developing objectives)	• Prepare facility • Assemble props and other enhancements • Brief participants • Conduct exercise	
Evaluation	• Select evaluation team leader • Develop evaluation methodology • Select and organize evaluation team • Train evaluators	• Observe assigned objectives • Document actions	• Assess achievement of objectives • Participate in postexercise meetings • Prepare evaluation report • Participate in follow-up activities

Good evaluation can help the organization identify:

■ Whether the exercise has achieved its objectives
■ Needed improvements in the EOP, procedures, or guidelines
■ Needed improvements in the emergency management system
■ Training and staffing deficiencies
■ Needed operations equipment
■ Need for continued exercising of the plan and the emergency management functions

If these goals are to be met, the evaluation approach must be systematic—methodical in procedure, thorough, and organized.

The Evaluation Team

In the early stages of exercise design, a number of people will be brought together into a design team, led by a design team leader or exercise director. One member of this team will become the evaluation team leader, or chief evaluator, and this person will in turn select the members of the evaluation team.

Team Structure

The size and composition of the evaluation team will depend on the type of exercise, its complexity, and the availability of people to serve. A small exercise with limited objectives, few participating organizations, and only a few locations might need only a team leader and three to six evaluators. A group this small would report directly to the team leader.

A large, full-scale exercise might require an evaluation director to supervise several team leaders at different sites, who in turn supervise several evaluators. With a team of this size, the various evaluators would be assigned to serve at selected sites, and a means of coordination or communication would be set up among them. An organizational chart would be helpful to keep lines of authority straight.

Role of the Team Leader — *What?* The evaluation team leader is primarily responsible for the evaluation methodology, for selecting and training the evaluation team, and for preparing the evaluation report.

Who? Ideally, the team leader should have experience in evaluation, management, exercise design or participation, and training or education. Normally, he or she should be a member of the design team. In some cases, it might be necessary to bring in someone from outside the design team, although that is usually unnecessary if a volunteer training and exercise officer is on the design team and is willing to serve. (Other design team members are likely to be too heavily involved in developing the exercise.)

When? Selecting the evaluation team leader as early in the design process as possible has several advantages:

- It will ensure that the evaluation becomes an integral part of the exercise development effort.
- It will maintain the integrity of the evaluation function and prevent it from overlapping with the control and simulation functions.
- It will ensure that at least one person can devote time and mental effort to the large task of evaluation.

Selection of Team Members — The evaluation team leader is responsible for selecting and training the evaluation team. The ideal evaluator has many skills and personal attributes. Although it may not be possible to find a person who has all of these characteristics, the team leader will be looking for the following:

Skills
- Appropriate technical expertise in evaluation
- Communication skills, both oral and written
- Organizational ability
- Ability to see the relationship between events and objectives
- Ability to adjust to rapidly changing situations

Attributes
- "People skills," sensitivity
- Objectivity
- Self-motivation
- Willingness to help
- Honesty and integrity (reports facts truthfully, keeps information confidential)
- Familiarity with the plan

Often an evaluation team can be assembled by using a little imagination and beating the bushes. Evaluators can be drawn from various sources, including:

- Neighboring jurisdictions
- Emergency services personnel who will not be playing in the exercise
- Professional evaluators
- State or federal personnel
- College or university faculty
- Public service organizations

Training of Team Members — The training of an evaluation team usually can be done in an orientation meeting. The nature and length of the meeting will depend on the experience and skill of the members. Most evaluators—no matter how experienced—will need information on the following:

- The exercise scenario
- Rules of play
- The objectives
- Evaluation requirements and procedures
- Evaluation forms

Inexperienced evaluators may need some practice drills. Evaluators from outside the organization/jurisdiction will need information about the organization/jurisdiction.

Try to be unobtrusive. It is well documented that the presence of an evaluator can affect the behavior of those being observed, possibly resulting in inaccurate data. Therefore, the evaluation team should plan ways to observe that are as unobtrusive as possible. Examples include:

- Being in position when the exercise begins so as not to attract attention
- Postponing taking notes for a few minutes until players get involved in the play and stop noticing them

Evaluation Methodology

Evaluation methodology is simply the procedures and strategy that are used to evaluate an exercise. The methodology includes:

- How the evaluation team will be structured
- Objectives to be measured
- Evaluation packet

Aspects of the team structure will have a significant impact on how the evaluation proceeds. Therefore, decisions such as the following are an important part of the methodology:

- Number of evaluators and their evaluation-related experience and knowledge
- Organization of evaluators at multiple sites (i.e., subteams)
- Lines of authority (e.g., evaluators, supervisors or team leaders, chief evaluator)
- Communication and coordination among team members

Evaluation Criteria

One of the first steps in developing the methodology is to determine the criteria that will be used to determine if the exercise is successful. These criteria are tied to the objectives and expected actions.

At the outset of exercise design, the general objectives are defined. Then, in developing the scenario, the general objectives are broken into smaller units—the expected actions. From the expected actions, specific points of review and other evaluation measures can be developed.

As discussed earlier in the chapter, the objectives must be stated clearly and precisely, describing actions that can be observed and measured. Using the SMART system will ensure that objectives are:

- Simple
- Measurable
- Achievable
- Realistic
- Task-oriented

You may wish to review the material now to refresh your memory about the important elements of good objectives.

Evaluation Packet

The evaluation packet, or evaluation plan, contains all plans for the collection of data, including objectives and points of review, checklists or other evaluation forms, and observation techniques.

Data can be collected by various means, such as evaluation forms, running written narrative, audiotape, or videotape. Each method has advantages and disadvantages, which should be considered in developing the methodology.

Suggested Evaluation Strategies

Plan the observation process. One approach is to use these four steps:

- Recall the specific objectives of the exercise, the detailed events, and the actions or decisions that they suggested.
- Identify the players expected to take the actions or make decisions as those who should be observed.
- Locate evaluators in a position to observe the players.
- Brief the evaluators on what actions or decisions are expected.

Provide points of review to guide evaluators. They make it possible to be very objective in collecting data. For example:

Objective: Demonstrate the adequacy of displays to support the emergency operations during the exercise.

Points of Review	Yes	No
Status boards available in facility	_____	_____
Status boards utilized	_____	_____
Status boards kept updated	_____	_____
Maps available	_____	_____
Maps up to date	_____	_____

Provide evaluation forms. They may include simple questionnaires, checklists, or rating sheets on which observations are recorded. They need not be complicated, but they must be objective, simple, and specific.

Key Event Monitoring

Most exercise scenarios include a number of events that are specifically designed to put stress on selected elements of the plan. These are termed *key events*. Evaluators should pay special attention to these events.

When a key event message is input, the evaluators monitor the participants' responses to the event. All responses should be noted on a key event response form. This form provides for multiple responses from several positions within the EOC as well as responses from outside the EOC.

Problem Log

The problem log allows participants, controllers, and simulators to document any observed action that may possibly create a problem. (While observing, it should be understood by everyone that what may seem to be a plan or procedure problem may actually be a participant or simulator error.)

These potential problems can then be analyzed after the exercise to determine which are serious enough to require corrective action and to determine their source(s)—plan, preparedness, training, or simulation.

Sample Forms

The following forms are provided here as job aids.

- Evaluator checklist
- Narrative summary form
- Key event response form
- Problem log
- Exercise debriefing log
- Exercise critique form

Postexercise Briefings

There are two types of postexercise meetings: the player debriefing and meetings of the evaluation team.

Player Debriefing

A short exercise debriefing should be conducted with the players immediately after the exercise. This debriefing gives them an opportunity to have their say about how things went, what they think should be changed, and commitments that they might make.

This is how a debriefing generally goes:

- The controller conducts the debriefing, beginning with a review of broad objectives and commenting generally on both successes and shortfalls.
- The controller asks for a brief response (about two minutes each) from each player.

As participants comment on their performance, the controller tries to maintain a balance between positive and negative comments, and encourages everyone to contribute. Comments during the debriefing should be recorded for inclusion in the after-action report. An exercise debriefing log may be used for this purpose. An example is provided at the end of this section.

The debriefing is for exercise participants. If evaluators wish to say a few words, they should concentrate on all the positive aspects of the exercise.

It is a good idea to prepare a simple questionnaire (such as the Exercise Critique Form in Figure 9.6) for participants to fill out after the exercise. People who hesitate to enter into group discussions will often respond to a questionnaire. One possible format is a set of objective questions requiring only a check mark response,

Please take a few minutes to fill out this form. Your opinions and suggestions will help us prepare better exercises in the future.

Please rate the overall exercise on the following scale.

| 1 | 2 | 3 | 4 | 5 | 6 | 7 | 8 | 9 |

Very Poor   Very Good

Compared to previous exercises, this one was:

| 1 | 2 | 3 | 4 | 5 | 6 | 7 | 8 | 9 |

Very Poor   Very Good

Did the exercise effectively simulate the emergency environment and emergency response activities? Yes _____ No _____

If no, briefly explain why:

Did the problems presented in the exercise adequately test readiness capability to implement the plan? Yes _____ No _____

If no, briefly explain why:

Figure 9.6 Exercise critique form.

along with some open-ended questions about the performance in general (e.g., What was good or bad about the notification procedure?).

Keep the Debriefing on Track

The purpose of the debriefing is to examine player performance. However, players will often want to critique the exercise itself: It was too long, too short, had too many or too few messages. The goal is to keep the players on track, focusing on performance insofar as possible. Explain that they will have an opportunity later to provide input into exercise design. Figure 9.6 is useful during the postexercise debriefing.

Evaluation Team Meetings

Evaluation team meetings are held to analyze the exercise and prepare the after-action report. Evaluation team meetings may include the following:

- A meeting of the evaluation team(s) shortly after the exercise to compare notes.
- A more formal meeting of the team a week or so later to analyze the findings and develop an accurate account of what worked and what did not. The team analyzes evaluation responses and any other data, and discusses how well each of the objectives was met.
- Additional meetings as needed to analyze data and prepare the report.

The exercise design team may join the evaluation team at one or more meetings to offer feedback and suggestions. The report should be prepared within one to three weeks after the exercise, while memories are still fresh.

After-Action Report

The findings of the evaluation team meetings are compiled in the after-action report, which documents the effectiveness of the exercise. It serves as the basis for planning future exercises, upgrading the EOP or contingency plan, and taking corrective action. The after-action report is the responsibility of the evaluation team leader or chief evaluator, working with the evaluation team.

Form

The report may take a variety of forms. For example, a small exercise may warrant only a brief summary of the minutes of the player debriefing, followed by a few recommendations. Sometimes a memo will do the job. For large exercises—particularly functional or full-scale exercises—the report should be specific and comprehensive.

Format

There is no set organizational plan for an after-action report. However, the topics listed in the following outline are usually covered.

1. Introduction
   a. (Main purpose of the report, why it is being submitted, preview of main topics, evaluation methodology used, and perhaps a general summary of main problems and recommendations.)
2. Statement of the Problem
   a. (Purpose of the exercise)
3. Exercise Summary
   a. Goals and objectives
   b. Preexercise activities
   c. Participants and agencies
   d. Description of exercise scenario
4. Accomplishments and shortfalls
   a. Evaluation group findings
   b. Summary of postexercise debriefing
5. Recommendations
   a. Training needs
   b. Changes in the emergency plan
   c. Other corrective actions

A copy of the after-action report should go the CEO in your jurisdiction and each participating entity.

Other Reporting Requirements

Other types of exercise reports are often required by agencies that mandate exercises. Often they are simple checklists that require less time to prepare than a full written report. Check with the appropriate regulatory agencies for specific reporting requirements.

Implementing Change

Recommendations for the future are the whole reason for conducting the exercise. The goals of an exercise are not achieved until the recommendations that come out of the evaluation are implemented. The purpose of the evaluation is to improve the emergency management plan and the organization's performance in carrying out that plan. This is most likely to occur if objectives relate clearly to emergency functions and the focus of the evaluation is on performance, not people. Specifically, the change effort centers on these issues:

- Are the procedures sound?
- Are resources sufficient to support the procedures?
- Are personnel adequately trained to follow the procedures and use resources?

Remember: We test plans, but we train people!

Enhancements

Sometimes creativity is the difference between an adequate exercise design and one that really gets the participants excited and involved. We'll now discuss the various ways in which you can enhance an exercise with equipment, displays, people, props, and other strategies.

Why Use Enhancements?

The point of an exercise is to simulate an emergency as realistically as the type of exercise warrants. The more realistic the scenario, the setting, the atmosphere, and the equipment and materials made available to the participants, the more likely it is that the players will get into the action and get the most out of it.

A variety of exercise enhancements can help achieve this realism. In a drill or full-scale exercise, the use of real equipment and actual locations is inherently realistic. Added touches—such as simulated victims with convincing mock injuries—can make a functional or full-scale exercise even more realistic. A tabletop or functional exercise must rely on materials and devices that you can bring into a room to make the exercise more realistic.

It's not necessary to spend a lot of money or energy to enhance realism. Resources can include ordinary items that are currently available in the operations center or command post, or common items that can be easily obtained.

Creativity

Creativity is the key to good enhancements. There are many low-cost creative approaches you can try. Here are some examples:

- Videotape simulated "news broadcasts" depicting the disaster and taped interviews with "victims." Show these prerecorded clips as part of the exercise.
- Audiotape news broadcasts and play them on the radio.
- Use makeup and props to simulate injuries.
- Use computers to chart plumes and provide data on river flows.
- If the telephone system is down as part of the exercise, leave the telephones in the EOC but don't connect them.

■ If the power is supposed to be out, then actually run a backup generator. Turn off the lights and computers if they're not hooked up to the generator. (Of course, this can disrupt the real office work flow, but it will result in a more realistic exercise.)

These are just a few examples for you to get your creativity flowing. We'll discuss some more ideas throughout the remainder of this chapter, and you will undoubtedly have some ideas of your own.

Communications Equipment

Various types of electronic equipment can be used to communicate the narrative and advance the scenario. Communications equipment can also be used to transmit some of the messages from simulators. In a highly simulated exercise, such as a complex functional exercise, try to transmit messages as you would in a real emergency—by phone, radio, and TV.

Take advantage of what's available. There are always financial limitations, so plan to make use of the communications equipment that your community normally has available during an emergency. Consider the following:

■ Landlines and cellular telephone
■ Radio phones and portable/handheld radios
■ Hotline dedicated phones
■ Military phone hookups
■ Citizens band (CB)
■ Teletype systems and fax machines
■ Amateur Radio Emergency Service (ARES) and Radio Amateur Civil Emergency Service (RACES)
■ Monitors/scanners
■ National Oceanic Atmospheric Administration (NOAA) Weather Radio
■ National Warning System (NAWAS)
■ Computerized radio packet

Visuals

No exercise is complete without a collection of visuals and displays: Maps, charts, status boards, black or white boards, chart paper and easels, and the like. Videotapes and slides, although more difficult to obtain, can also enhance the realism of your exercise.

Maps

Maps provide context and detail for a scenario. Because they are essential to the handling of an actual disaster, they are necessary for all types of exercises. Even in an orientation or a tabletop exercise, maps provide useful information and give players a clearer picture of the simulated event. For example, they may be used to position equipment or to determine the nearest facilities for resource deployment.

Maps may be reproduced on paper for individual use or displayed on a wall. Acetate overlays make it possible to mark off areas or monitoring points and to reuse the map. The types and number of maps required depend on the exercise type and the hazard being tested. Following is a partial list of maps that you may find useful.

- City street maps
- County street maps
- Subdivision maps
- Sewer maps (mains/facilities)
- Water maps (mains/facilities)
- Electric maps (lines/facilities)
- Gas maps (lines/facilities)
- Flood plain maps
- Contour maps
- Police and fire district maps
- Center city (downtown) maps
- Facility maps (plant layout, rail yards, etc.)
- Weather maps

Charts Used in Exercises

The accumulation and sharing of information is an important operations center function. To ensure coordinated and timely emergency response, visual displays allow everyone to comprehend quickly what actions have been taken and what resources and personnel are available.

Display needs will vary with the nature and scope of the exercise, but the charts listed in Table 9.7 should be considered. Some of the displays are for use of the exercise staff only.

Videotapes and Slides

Videotapes can provide a very realistic presentation of an emergency that can be used to introduce the narrative or to give updates. News reports and interviews with politicians and the public can be prerecorded to lend realism. Slides can be used for some of the same purposes.

Computer Resources

Many communities have their resources on a computerized inventory list. Where available, this inventory should be used during the conduct of exercises to track resources.

Table 9.7   Types of Charts

Type of Chart	Description
Problem and event log	• Large-events display board for posting major events • Should be available for all in the EOC to review; also useful for EOC shift-change briefings • All major problems that are reported are entered on the log as they are received • May be divided into columns: nature of problem, problem number, assignment, response, remarks
Damage assessment chart	• Divided into columns: areas reporting damage, time of report, extent of damage
Facility charts and status boards	• Used to track facilities involved in the exercise so that participants are aware of available resources (companion maps are useful); examples of facility charts: • Hospitals: beds available, blood and other supply needs, personnel • Congregate care facilities (e.g., those run by Red Cross or social service agencies): space available; status of food, water, bedding, medical stocks • Law enforcement resources: numbers and locations of sworn, reserve, and auxiliary personnel; status of mutual aid units • Fire resources: deployment and availability of fire units, status of fire mutual aid forms
Organization charts	• Useful for staff as a means of anticipating what agencies should be coordinating with or reporting to other agencies • Optional
Master scenario of events list	• Mainly for controller's use, to keep exercise on schedule; should not be seen by players • Contains detailed sequence of events developed as part of scenario
Simulation plotting map	• Used by controller and simulators • Depicts prescribed input exercise information • Coded markers may be used to depict actions taken by various organizations (police, fire, medical/health, public works, utilities, Red Cross/voluntary agencies)

Miscellaneous Equipment and Supplies

Sometimes exercise planners are so focused on the dramatic aspects of the exercise that they overlook common equipment and materials. Many of these items, available in most office stores at little cost, are indispensable. Here are some equipment and supplies to consider:

- Projector for overhead transparencies or slides
- Copy machine
- Portable radio
- Pagers and cellular phones
- Public address system
- Pens, pencils, and markers
- Chart paper
- Telephone books and directories
- Local and state contact lists
- EOC phone directory
- List of cellular phone numbers
- Name tags
- Resource lists
- Financial cost report form

People and Props

In a full-scale exercise, the sense of reality occurs through the use of actual equipment in the actual setting. In this setting, fire trucks and the local airport would be considered types of enhancements.

Designers of full-scale exercises also bring in real people or props to enhance the realism. For example, an exercise simulating a hotel fire could use the following:

- People playing the part of victims, made up to appear injured
- Mannequins to represent the dead (or victims trapped under heavy beams)
- Fake smoke
- Burnt boards and beams strategically placed at the event site
- A contained fire that the fire department would be required to extinguish

Some jurisdictions use considerable ingenuity in these matters—creating model cities to use in tabletop exercises or obtaining mannequins to substitute for humans in dangerous situations (e.g., trapped under a beam). Such enhancements are limited only by budget, safety considerations, and the imagination. So let your creativity work to enhance the exercise that you are planning.

Enhancement Resources

One of the problems that designers often face is how to obtain materials, people, and equipment that lend realism without breaking a very small budget. Consider soliciting volunteers and donations from other agencies in the community. Many are civic-minded and are happy to help out by lending equipment or providing volunteers to serve as victims. Some potential resources include:

- Hospitals
- Schools and colleges
- Business and industry
- Chemical Transportation Emergency Center (CHEMTREC)
- Boy and Girl Scouts
- Search and rescue volunteers
- Agencies
- Railroads
- Salvation Army
- Public transportation
- Fire departments
- Police departments
- Chemical companies
- Amateur radio clubs
- Religious organizations
- National Guard/military
- American Red Cross
- Service organizations (Elks, Lions, Rotary, etc.)

Costs and Liability

In obtaining resources, it is important to consider costs (both initial and hidden) and the potential for liability. For example, consider:

- Person hours expended in obtaining and returning equipment or materials.
- Potential for damage or replacement costs.
- Arrangements for timely returns when items are borrowed or volunteers are "on loan." (Too often "victims" have been left at the scene of the emergency site or the hospital because transportation plans were not made.)
- Safety and liability concerns when using people and equipment. Be sure that provisions are in place for liability insurance and equipment replacement.

Enhancement Logistics

In planning for exercises, someone needs to take responsibility for managing the logistics related to enhancements. Be sure that the following questions are answered:

- How will enhancements be used?
- Where will props be placed?
- Who will be in charge of props?
- How will people and props be picked up, transported, and returned?
- What kind of return policy can be worked out for borrowed materials and equipment?
- In what condition must equipment be returned?
- Who will clean it?
- Is normal wear a concern?

Conclusion

Exercises are conducted to evaluate a jurisdiction's capability to execute one or more portions of its response plan or contingency plan. Exercises can be used to provide individual training and improve the emergency management system. Reasons to perform exercises include:

- Testing and evaluating plans, policies, and procedures
- Revealing planning weaknesses and resource gaps
- Improving individual performance, and organizational coordination and communications
- Training personnel, and clarifying roles and responsibilities
- Gaining program recognition
- Satisfying regulatory requirements

A comprehensive exercise program is made up of progressively complex exercises, each one building on the previous one, until the exercises are as close to reality as possible. The program must be carefully planned to achieve identified goals and should involve a wide range of organizations in its planning and execution.

Building an exercise program is a multiorganization team effort that includes the following:

- Analysis of capabilities and costs
- Goal setting
- Development of a long-term plan
- Scheduling of tasks
- Public relations efforts

The process of creating and staging an exercise includes a lengthy sequence of tasks that occurs in three phases: before, during, and after the exercise. Some of the tasks fall under the heading of design and some are part of evaluation.

A simple way of viewing the exercise process is the sequence of five major task accomplishments:

1. Establishing the base
2. Exercise development
3. Exercise conduct
4. Exercise critique and evaluation
5. Exercise follow-up

Tabletop, functional, and full-scale exercises are based on a design process that includes eight steps that are generally applicable regardless of the type of exercise. Each type of exercise has some special considerations in how these steps are applied.

Outputs from the design process are pulled together in the master of scenario events list (MSEL), a chart that the controller and simulators can use in keeping the exercise on track and on schedule.

The tabletop exercise is essentially a group brainstorming session centered on a scenario narrative and problem statements or messages that are presented to members of the group. The format is informal, and the exercise is self-evaluated by the participants.

The functional exercise usually takes place in the operating center and involves policy makers and decision makers. It uses an event scenario to test multiple functions or organizations, emphasizing coordination and communication.

The full-scale exercise combines the interactivity of the functional exercise with a field element and requires the coordination of the efforts of several organizations. It differs from a drill in that a drill focuses on a single operation and exercises only one organization.

For an exercise to be useful, it must be accompanied by an evaluation—less formal for the tabletop, structured for the functional and full-scale. Good evaluations can help the organization identify the following:

- Whether the exercise has achieved its objectives
- Needed improvements in plans, procedures, or guidelines—or the emergency management system as a whole
- Training and staffing deficiencies
- Equipment needs
- Need for additional exercising
- The evaluation team leader—usually drawn from the design team—is responsible for evaluation methodology, selection and training of the evaluation team, and report preparation.

Postexercise meetings include the player debriefing and meetings of the evaluation team to analyze the results and develop the after-action report, which should describe the purpose of the exercise and address goals, objectives, preexercise activities, participants, scenario, accomplishments and shortfalls, and recommendations.

Chapter 10

EOC Management and Operations

Lucien Canton and Nicholas Staikos

Contents

Introduction

It is a common mistake to confuse the emergency operations center (EOC) with the tasks that are performed in the EOC and to forget that the EOC is a physical location that generates its own demands. For the EOC team to perform effectively, the physical and organizational demands of the EOC as a facility must be met. This EOC management is distinct from the operational management of the incident.

EOC management can be roughly divided into two main categories: facility management and operational management. Facility management is similar to the activities that take place within any facility. This involves dealing with the physical plant, technology systems, and support services needed to support activities within the EOC. Operational management pertains to the systems and procedures put in place to allow the EOC team to operate efficiently. This pertains to the procedures for performing common tasks and operating EOC equipment. These two components work together to ensure that the EOC team is free to focus on the incident with minimal disruption from the environment in which the team is operating.

Facility Management

EOC facility management shares many commonalities with that found in any major office building. However, there are added complexities that make EOCs unique. Where the typical office building operates for eight to twelve hours a day, an EOC must be capable of twenty-four hour operation. This means that there is no downtime for maintenance or support services, and these services must be provided in a way that does not have an impact on operations. In addition, increased security during operations can limit access for support staff or contractors if there is no prior coordination. The EOC also has a considerable number of parallel systems and complex communications systems and may have unique design features such as HEPA filters or overpressure systems to provide security against chemical and biological attack. In some cases, the EOC may also provide living quarters for staff for an extended period of time.

This level of complexity for EOC systems has implications for day-to-day activities as well. Since an EOC must be ready for activation within a short period, there is no latitude for a lengthy start up period while batteries are charged, software upgraded, or systems checked. This means that a program for ongoing maintenance of EOC systems must be in place.

The EOC as a facility comprises a number of systems. The most obvious are, of course, the environmental, life safety, and utility systems. Even here, though, the EOC is different from a typical office building. Where "emergency power" in an office building means that life safety systems continue to receive power, the EOC's requirement for continuous operation under all conditions may demand alternate commercial power feeds, multiple generators, and the capacity to add external generators.

The same is true for communications and information technology systems. EOCs require multiple parallel systems, each with unique requirements. In some cases, these systems cannot operate simultaneously.

In addition to technical support, the EOC also places demands on support services. For example, twenty-four hour operation will increase the need for janitorial service and garbage pickup. Additional materials may need to be ordered and delivered. Contractors may be brought in to provide food for EOC staff.

One critical service that must be coordinated is that of security. A fully activated EOC is a focal point for the media and members of the public and could be viewed as a potential target. While day-to-day security may be adequate, during activation there will be a need for additional security to control access, provide escorts, and conduct patrols. Security responsibilities may include controlling and protecting parking areas or adjacent staging areas.

Unfortunately, facility management is a multidiscipline task performed by many different individuals. The typical jurisdiction will have the physical plant serviced by one department while another oversees information technology systems. Radio systems may be handled by a separate communications department, while vendor services are overseen by still another department and security by another. It is rare to find a single person charged with looking at the EOC facility systematically. This creates problems during day-to-day operations, but it can prove devastating during actual EOC activation. With no go-to guy for facility issues, problems begin to accumulate and distract the EOC team from its operational responsibilities.

For this reason, emergency managers should consider creating an EOC coordinator position that serves as the central point of coordination for the vendors and teams supporting the EOC facility. This team approach is consistent with how the EOC team routinely operates; the difference is that this team is focused on internal support to the EOC rather than support to the overall operation.

Operational Management

There is often an assumption that when staff report to the EOC, the only thing they really need to function efficiently is the emergency operations plan (EOP). After all, the EOP contains a substantial amount of information, policies, and procedures focused on coordinating disaster operations. Coupled with internal department plans and standard operating procedures (SOPs), the EOP should be sufficient to guide operations.

This may be true in terms of the disaster operations but is not true for the EOC as a facility. Consultant Art Botterell's Third Law of Emergency Management states that "no matter who you train, someone else will show up." EOC planners need to assume that a significant percentage of the EOC team will be coming to the EOC for the first time. Other members of the team may not have been in the EOC for some time and forgotten how things work.

As an example, consider simple tasks related to information management. How will a team member receive an e-mail? How can he or she direct it to print to the closest printer? What's the EOC fax number? These are simple questions to answer, but if fifty people are asking them at the same time, it can considerably delay operations. Even as simple a question as "How do I turn on the lights?" may create problems if the first person to arrive at the EOC has never been there.

In addition to these more technical issues, there are a number of tasks that need to be performed to support the EOC team. An example of these is a process for activating the EOC. This process should address basic questions such as who has authority to activate the EOC and under what conditions, but it should also describe in detail the procedure for notifying the EOC team.

Another overlooked process involves the procedure for the initial setup of the EOC upon activation. Something as trivial as not knowing who has the keys or how to turn on a copier can cause considerable problems. Many EOCs are dual-use facilities that require the first arriving staff members to unlock containers, plug in telephones, and set up laptop computers. Assuming that the first person at the EOC will be one of the three or four staff members that know how to set up the facility is a fundamental mistake.

Once operations are concluded, there are things that need to be done to close out the current operation and prepare the facility for future operations. This involves procedural tasks such as archiving files and preparing after-action reports as well as facility-related tasks such as clean up, repairing or performing maintenance on equipment, and reordering supplies. The tendency will be for the last remaining team members to quickly wrap things up and go home, and important tasks can be overlooked without a detailed deactivation process.

This need for structure and organization reinforces the need for standard operating procedures that are focused on the EOC rather than on external operations. The EOC SOP should be written from the perspective of the first-time user and, as much as possible, reduce procedures to short checklists. The checklists should be detailed and task-oriented. Among the items that could be included in the SOP are the following:

- Activation procedures
- Notification procedures
- Setup procedures
- Procedures for using communications and information technology systems
- Procedures for obtaining additional supplies and services
- Security procedures
- Deactivation procedures

One other technique to consider is the posting of key information from the SOP on wall displays in the EOC for ready reference. This could include procedures for transferring calls, fax numbers, printer addresses, and other information that supports operations. Similarly, instructions for use can be posted by copiers, fax machines, and so forth. Remember that staff using the EOC will be under pressure and may be seeing the EOC equipment for the first time. Botterell's First Law of Emergency Management cautions that "stress makes you stupid."

Organizing for EOC Management

Something that is everyone's responsibility is in reality no one's responsibility. An important first step in developing a plan for EOC management is to identify who will be responsible for overall planning and who will serve as the EOC coordinator during activation. These two positions do not necessarily need to be the same person.

Planning for EOC management is no different from planning for operations. One begins by identifying key players and stakeholders, and developing a working group to guide the work. As EOC systems and procedures are developed, they should be tested through exercise, preferably by being integrated into regular EOC functional exercises. While this seems obvious, relatively little planning is traditionally done in the area of EOC management; most organizations focus all their efforts on operations planning. This is the equivalent of assuming that normal departmental functions will be available during a disaster in the absence of a continuity of operations plan. It's a very risky assumption that could result in operational failure.

The following are key points to consider when planning for EOC management:

■ Identify a lead person with overall responsibility for EOC management planning
■ Identify key players and stakeholders, and form a working group
■ Develop standard operating procedures for common EOC functions
■ Develop standby contracts for increased or additional services
■ Coordinate supporting operations such as security and support services
■ During activation, consider forming an EOC support team under a single coordinator
■ Use a standard deactivation process that prepares the EOC for immediate reactivation before concluding operations

Good EOC management can significantly reduce confusion and stress on the EOC team, allowing the team to focus on operations rather than being distracted by their operational environment. It requires preplanning in a manner similar to operations and an ability to forecast the needs of the EOC team. In the end, it may well determine the success or failure of the EOC operation.

Operations Room Design
Origins

The operations room, where internal and external responders report, is the nerve center of today's EOC. Its evolution, like the field of emergency management, is ongoing, driven by the constantly changing technological landscape and by the adoption of new practices derived from lessons learned. It was initially conceived as a space where key public and private agency representatives came together for the collection, evaluation, and dissemination of information. These multipurposed spaces were born out of the context of civil defense. They were fairly similar in layout and typically created with a bunker mentality as survivability was paramount.

The accommodations were spartan in nature and the facility was used for other purposes until escalating threat levels warranted activation. While their principal function was to provide the responding agencies a seemingly protected place to maintain communications with their respective organizations' operational structures, they were also viewed as a place of refuge for governing authorities, thus ensuring the continuity of government. It wasn't unusual to find these centers located in the lower basement levels of a municipal building or at times as a stand-alone underground facility. Interestingly these presumed-to-be-well-protected locations were held hostage by external events outside of their control such as plumbing failures from above or flooding via backed up drainage systems from below. Further, by being collocated with many other users, any form of building evacuation such as a fire alarm or other such alert would mandate exit from the occupied space not to mention the impact of sprinkler discharge seeking the lowest level.

As the Cold War tensions defused, officials soon came to the realization that the threats from natural disasters would be more likely to occur and could have significant consequences on the day-to-day function of government as well as affecting the lives and welfare of a large proportion of the jurisdictions' population. This growing awareness produced the need for enlarged staffs to administer recovery programs as well as focused support during the crisis and drove the need for more capable facilities.

The September 11 attacks shifted emergency management's focus from an all-hazards approach to one that was biased toward homeland security and counterterrorism. The concern over the short-sightedness in this shift in emphasis was raised by the emergency management community. Then, the impact of Hurricane Katrina and the other storms of the period reminded everyone that emergency management and response demands a broad base of preparedness found in an all-hazards approach.

This transitional awareness is helping drive a new mindset among community leaders that EOCs and their operations centers should not only ensure survivability of government but facilitate continuity of operations for the private sector by striving to quickly return to normalcy. Our leaders recognize that this is even more important now that globalization and interdependency are no longer academic concepts but a reality. The linkage between suppliers, manufacturers, producers, and consumers is one of the key threads that binds a nation together. Having a robust and strong economic base has come to represent a key component of a country's strength. As a result, the EOC and its operation's room play a pivotal role in managing a crisis, and the EOC has had to became a sophisticated communications hub to fulfill its mission during the cycle of mitigation, preparation, response, and recovery rather than just a command and control center during a crisis.

Today's Focus

Regardless of the scale of an emergency, the success of a coordinated response almost always depends on several key factors:

1. Redundant and interoperable communications systems, which are balanced vertically and horizontally
2. Comprehensive ability to quickly determine and coordinate asset utilization
3. Organizational flexibility to accommodate a variety of responding entities, which are driven by event type
4. In-depth situational awareness
5. Access to all supporting resources to formulate alternative response scenarios

One must also recognize that because of the speed at which events can unfold and the need to engage a myriad of supporting players, a well-crafted plan should be in place. To be effective it should target the most likely types of events for the locale and be tuned as a result of multiagency exercises to be fully effective. The plan—the tools—are the baseline of preparedness, for as we have often seen, events never follow the script and scenario adaptation will be necessary.

Another reality that our political establishment can sometimes forget is that all disasters are local. Even during a widening crisis, organizational effectiveness starts with the local responders and then gradually draws upon the next level of support. Having said this, we have all seen situations where the local entity failed to quickly recognize that events were spiraling beyond their capabilities and failed to request support quickly enough. This is where the value of the EOC's operations room is leveraged. For while incident command is focusing on the immediate issues on the ground, the professionals in the EOC with a theater-wide view can implement and guide a strategic response measured by the needs coming from the field.

Additionally, we must further recognize that staffing realities dictate that no matter the size of the jurisdiction, support during the initial stages of an event will come from a 24/7 watch component. This on-duty team acts as the trip wire providing the vital linkages in the response chain. They will do so until such time as the facility reaches full staffing wherein interface with each supporting agency's organization will be handled by their responding entity. This watch staff, therefore, will need to have all the skills and tools to capably manage the initial response until such time as the appropriate representatives of the activated agencies are dispatched to the EOC to provide real-time coordination within their respective infrastructure.

Can Organizational Structure Impact Room Design?

The short answer to this question is a definite yes. Our experience has shown that the management structure of the responsible agency most definitely influences the style of layout for the OPS room. Predictably, it will tilt either toward a C2 (command and control) or to a C4 (which I prefer to refer to as communicate, collaborate, coordinate, and, to a more limited degree, control) setup. The determining factor for this bias is quite often dependent on whether the agency has a public safety or emergency management heritage. In public safety environments where there is a clear chain of command as in the military, room layout will focus its attention on either a common information display wall or command structure. Those agencies from an emergency management lineage will tend to tilt toward a collaborative or clustered environment. In many jurisdictions hybrid models have begun to be implemented.

Regardless of the layout, the reality is that a fundamental shift has occurred and the modern EOC's success will be dependent on its ability to foster the coordination and disseminating of information to the appropriate consumers. These activities will be coordinated through the event managers and the emergency support function (ESF) personnel who are positioned in the operations room. And even as advancements in technology lend credibility to the notion that the future will be in the creation of a virtual EOC, recent events continue to suggest that face-to-face collaboration is a more efficient and effective form of problem solving. This reality, I believe, extends to all levels of government.

Design's Role in Supporting the Evolving Mission

One must keep in mind that the development of a comprehensive design for an EOC and its operations room must take into account a wide range of considerations that will affect internal and external features. These items encompass issues such as responder accessibility, hazard zone proximity, availability of redundant services, maintenance of secure operations, leveraging natural hardening through siting to physical hardening of both structures and systems. All of these factors as well as others must be considered when developing the design criteria and programmatic requirements for the facility. The degree to which a facility is hardened is largely influenced by the risk analysis developed from the threat and hazard assessment. The focus of this discussion will be on the various design options for the internal organization of the operations room of the modern EOC to create a well-conceived and functional center.

Management studies have shown that design does indeed have a significant impact on workplace effectiveness. Repeatedly, environmental comfort for both the physical as well as psychological needs of the occupants has been shown to play a significant role in mitigating the detrimental affects of high stress, which accompany crisis situations. Of the many factors that form the basis of an integrated design, the following are a few key components:

Ergonomics
- Console design
- Visual display design
- Seating comfort
- Technology integration
- Adaptability to a diversity of user body types

Environmental comfort
- Variable glare-free lighting control
- Acoustical control
- Thermal comfort and control

Space allocation
- Operator positions
- Supporting services
- Breakout areas
- Policy room
- Strategic response planning
- Quiet rooms
- Resource management
- Extended stay accommodation
- Self-sufficiency

Circulation and access control
- Hierarchical circulation system
- Electronics used to augment physical security design
- Layout supports work flow

Relationship to support spaces
- Ease of accessibility
- Ability to be serviced while in operation

Sustainable utility systems
- Redundant services
- Diverse routing
- Resupply capability
- Flexible cable management system

By properly considering these as well as other requirements, the designers of the workplace environment will have a significant impact on the operational effectiveness, thus shaping the quality of an entity's response. When implemented in the appropriate manner, these features and concepts will become transparent to the user as they will not be a source of discomfort. This will improve the level of performance dramatically as fatigue and frustration play a prominent role in degrading operational effectiveness.

The impact of design becomes even more apparent as the scale of an event escalates as these challenges become more complex due to the impact of their potential consequences. Time and time again, we read of the failings of response efforts, such as with Hurricane Katrina and in the earthquake in Haiti, when events of a catastrophic nature can overwhelm the system. Logistical entanglements, lack of communication, or conflicting requests produce chaos and unacceptable results. Make no mistake, facilities, whether physical or virtual, without a well-coordinated plan along with effective communications and logistical support will not produce the needed result.

Fundamentals

Each jurisdiction needs to have a solution tailored to its needs. The following narrative and accompanying diagrams illustrate several of the fundamental ways in which the focal point of an EOC, the operations room, can be configured to optimize the effectiveness of a jurisdiction's response. As the reader reviews the concepts, it is important to keep in mind that the efficiency of operation improves when the functional space is purposefully built yet affords the flexibility to adjust for refinement of operation. At times this may seem to be an unreachable goal, but it is achievable. This discussion deals primarily with the space and big-picture issues of technology integration and not its deployment nor optimal position assignment for the responding entities. Additionally, each of the plans presented can support the requirements and goals of the National Incident Management System (NIMS), some better than others.

One should also remember that many of these requirements, including situational awareness, asset control, and collaborative problem solving, are scalable to all jurisdictional levels. Further, the need for this capability and attendant sophistication increases as the jurisdictional landscape and physical area encompassed grow due to increased political complexity. Regardless of the many variations of layout currently in use, the design of the operations room can be characterized by six basic configurations, each of which can be applied to all levels of response. For the purposes of this analysis, we will describe them as follows:

- Traditional multipurpose
- Cubicle cluster
- Horseshoe
- Stadium/theater
- Collaboration pods—theater style
- Iris

Even virtual centers will utilize similar organizational structures for the network control center.

Although they may be referred to by other names, these are the basic configurations for today's modern operations room. As previously mentioned, the preference of one design form over another is often influenced by the branch of government in which the agency finds itself. This in turn influences the management style of the leadership and whether the organization's roots are from public safety, which favors a C2 arrangement, or from emergency management, which tilts toward a C4. Of course this doesn't mean that you can't have a collaborative environment with a central focus context.

In addition to the design of the internal arrangement of the operations room, it is essential that a well-thought-out concept of the supporting spaces for functions such as policy making, strategic planning, breakout rooms, quiet rooms, sleeping areas, equipment rooms, locker rooms, and break rooms be implemented. The interrelationship of these components plays a significant role in establishing an effective center and has great influence on the success of the center during periods of activation.

Review of Layouts

Multipurpose

The multipurpose layout was traditionally used in smaller jurisdictions where dedicated space for response activities was not available. This template featured a simple room with a flat floor, which could serve a multiplicity of uses from conference space to community meetings (Figure 10.1). Because of its multipurpose nature, conversion to a full-fledged OPS room/EOC required setup time for furniture configuration, technology deployment, communications installation and so forth.

Positives
1. Multipurposed
2. Flexible configuration

Negatives
1. Time required for physical setup
2. Technology deployment can be challenging; use of floor boxes can reduce set up time
3. Need storage space to store supporting equipment
4. Challenges in lighting control when layout changes
5. With flexibility comes lack of focus
6. Acoustics generally are substandard

Cluster/Pod Example

Figure 10.2 shows a cluster/pod example.

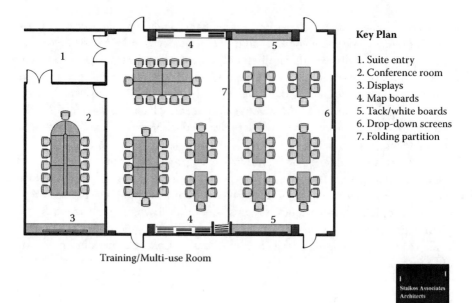

Key Plan

1. Suite entry
2. Conference room
3. Displays
4. Map boards
5. Tack/white boards
6. Drop-down screens
7. Folding partition

Training/Multi-use Room

Figure 10.1   Traditional multipurpose.

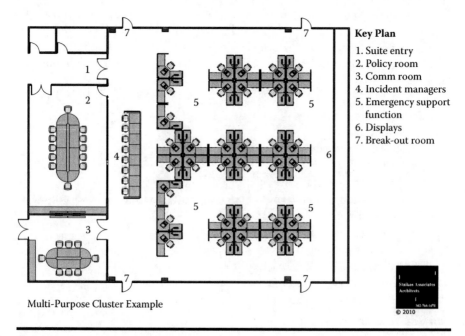

Key Plan

1. Suite entry
2. Policy room
3. Comm room
4. Incident managers
5. Emergency support
   function
6. Displays
7. Break-out room

Multi-Purpose Cluster Example

Figure 10.2   Cluster/pod example.

Positives
1. Allows for reconfiguration to enhance collaboration of responding entities
2. Allows for reassignment of cubes to entities having a higher level of interaction
3. Demands uniformity of technology to be conveniently flexible
4. Allows adoption of a hoteling concept with user logins to redefine occupant
5. Can adjust to variable numbers of related responder personnel

Negatives
1. Does not provide a central focus for disseminated or displayed information
2. On-the-fly reconfiguration is not as simple as it would seem and therefore not typically done until after an event, on lessons learned basis prior to the next event
3. Reconfiguration based upon specific event experience may not apply to a different type of activation
4. Use of cube positions even with seated height walls are harder to reorient than simple desking
5. Conflict between dedicated services and storage capability versus flexibility

Horseshoe

The development of the horseshoe layout allows the participants to view commonly displayed information yet maintain eye contact with their fellow responders (Figure 10.3). This arrangement is typically geared to a smaller room so that some direct conversation can occur across the room. When arranged on a stepped floor, supporting associates of the principals seated at the primary table can be positioned in close proximity.

Positives
1. In a smaller room environment, eye and voice contact can be maintained so that each speaker can follow the body language of the other principal
2. Opportunity for tiered or level OPS floor
3. Opportunity for centrally focused display wall
4. Allows for OPS floor breakout area or miniconference room, close at hand supervision
5. Provides room for supporting staff providing back up to the principals

Negatives
1. Has a tendency to encourage and generate increase noise levels if multiple conversations are occurring simultaneously
2. Doesn't effectively allow for cross-agency collaboration or subgroup collaboration as it must occur outside the room
3. Shape limits optimal or preferred sight lines
4. Limited number of ESF positions
5. Not ideal for small-group collaboration
6. OPS room becomes longer, narrower as positions increase

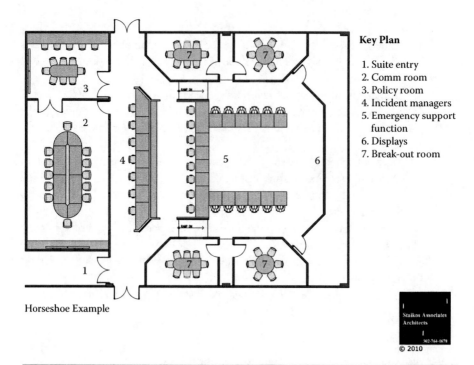

Horseshoe Example

Key Plan

1. Suite entry
2. Comm room
3. Policy room
4. Incident managers
5. Emergency support function
6. Displays
7. Break-out room

Figure 10.3   Horseshoe.

Stadium/Theater

The stadium/theater layout became popular with the advent of space missions wherein multiple activities could be commonly displayed for use by many participants. The rooms are generally stepped with tiers of consoles to allow unobstructed sight lines to displayed information (Figure 10.4). It fits a command and control model influenced by military models.

Positives
1. Visual focus of room allows all participants to share in tangential information used on an as-needed basis; messaging and pictorial information displays are easily viewed
2. Minimizes excess noise generation due to people asking what's going on
3. Allows attention to be focused on the speaker who may address the group
4. Incident managers positioned to oversee activity on OPS floor
5. Technology is fully deployed and tested so that startup is almost immediate
6. Can be effectively used for multipurpose training without reorganization

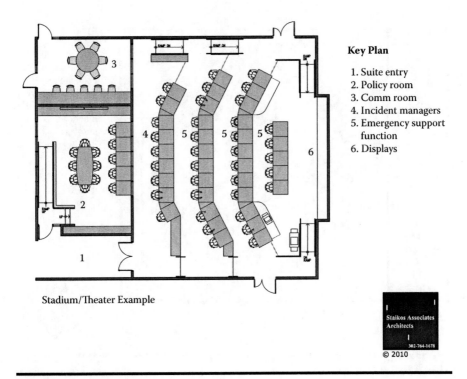

Stadium/Theater Example

Key Plan

1. Suite entry
2. Policy room
3. Comm room
4. Incident managers
5. Emergency support function
6. Displays

Figure 10.4 Stadium.

7. Adjacent or on OPS floor breakout area or miniconference room
8. Tiered concept allows for executive management to have overlook but be off floor
9. Allows for designed acoustical and light control methods to be in place

Negatives

1. Dedicated space with limited multipurpose use
2. Although positions can be reassigned easily, it does not readily lend itself to podlike clustering or grouping of responding agencies
3. Display size must increase with overall depth of room
4. Requires buffer area at front of room to maintain optimum viewing angles
5. Room proportions dictated by display characteristics
6. Large group collaboration must occur in break rooms

Collaboration Pods/Theater Variation

The pods/theater layout allows for the establishment clustering of ESF functions while at the same time providing common focus to displayed information. Configuration establishes subgroup principals allowing direct voice and visual

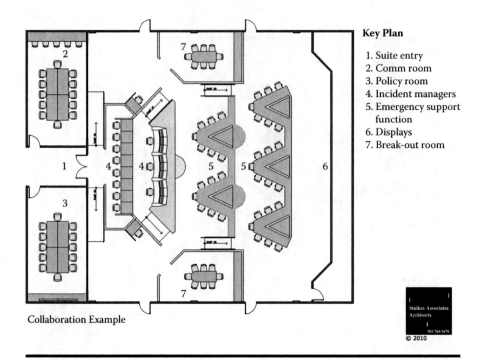

Key Plan

1. Suite entry
2. Comm room
3. Policy room
4. Incident managers
5. Emergency support function
6. Displays
7. Break-out room

Collaboration Example

© 2010

Figure 10.5 Collaborative.

interaction of the principals while still at their positions, allowing assimilation of ongoing event data (Figure 10.5).

Positives
1. Subgrouping or ESF clusters work well for seven to eleven participants as speaking levels can be held in check.
2. Provides all participants with front of the room focus to view displayed information
3. Stepped configuration allows for clear sight lines
4. Subgroup principals have direct eye contact with team members and information displays
5. Room for on-the-floor huddles between ESF subgroups
6. Breakout rooms can be located on either side as in theater/stadium layout

Negatives
1. Viewing angles are compromised for a portion of the group
2. Doesn't easily lend itself to changing ESF populations without migrating across POD structure
3. As depth of room increases, display surfaces need to enlarge, affecting optimal sight lines for those closer to display

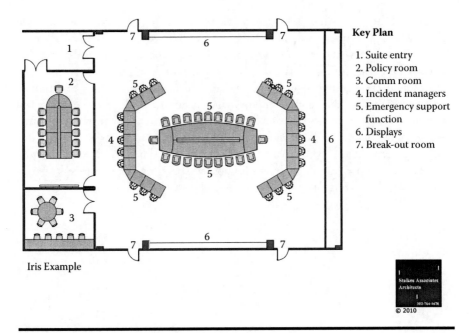

Key Plan

1. Suite entry
2. Policy room
3. Comm room
4. Incident managers
5. Emergency support function
6. Displays
7. Break-out room

Iris Example

© 2010

Figure 10.6   Iris.

Iris Example

The iris layout is patterned after a nonhierarchical structure where there is no self-evident position of room leadership (Figure 10.6). This deference to equality can be seen in the UN Security Council layout.

Positives
  1. Inward focus not hierarchical
  2. Acceptable sight lines but for a limited number of participants
  3. Breakout clusters/teaming outside OPS room

Negatives
  1. Poor sight-line angles due to need to place display walls at an elevated height
  2. Limited single point overview/observation opportunities caused by multidisplay walls
  3. Single level/flat OPS floor degrading sight lines
  4. Less flexibility for routine (nonactivation time) activities

Summary

As the reader well understands, the design of today's emergency operations center is a complex problem requiring an understanding of functional requirements, operational methodologies, and systems technologies. Regardless of the direction taken, it is a workplace that must be adaptable to meet ever-changing systems, management styles, and mission objectives. This can be accomplished through a concerted team-planning effort by fully engaging the emergency manager, architectural and engineering design professional, and facilities management team to produce a flexible, reliable, and maintainable complex. When this collaboration is implemented at the outset with a broad-based needs assessment, programming, team-oriented design, and continuing through to live performance evaluation, one can be assured that everyone's voice has been heard. Additionally, if compromise is dictated, the opportunity for buy-in is facilitated as all will be aware of the constraints, which will enable acceptance and produce a successful result.

Chapter 11

ICS/EOC Interface

Michael Fagel

Contents

After-action reports and studies of catastrophic disasters have identified the need to dispense additional information in emergency operations center (EOC) management. Although terrorist incidents are much less common, the same need holds true for these events as well. Major emergencies resulting from terrorist incidents can spread across multiple jurisdictions and may require a large-scale response. Emergency services must be ready and communication systems must be in place.

Every community, large or small, urban or rural, will be able to improve its ability to centralize its flow of information during an emergency by establishing an EOC. The key to a community's disaster planning, response, and recovery lies in the EOC. In fact, the EOC is crucial to saving lives and reducing property damage. To ensure that effective coordination takes place during all phases of emergency management, the emergency managers will work closely in a team environment with other EOC personnel, elected officials, and members of private-sector groups.

EOC Management and Operations: Responsibilities

Many of the emergency-related duties or tasks that need to be accomplished at the EOC are coordinated by the emergency manager, but may be carried out by other individuals designated by the emergency manager. This person is often given primary responsibility for duties related to EOC management and operations. Other designated EOC personnel may perform non-emergency-related tasks.

Duties and responsibilities for EOC management and operations may vary according to jurisdiction; however, there are core tasks that any designated person will be required to perform. Some of these tasks may arise before the EOC is activated, whereas others are ongoing. When the EOC is activated, additional responsibilities involve the direction, control, and coordination of numerous activities that develop in an emergency situation. Figure 11.1 lists the typical duties for EOC management.

Foundations for Establishing the Emergency Operations Center

A jurisdiction's emergency operations plan (EOP) and hazard vulnerability should be in place prior to the establishment of the EOC, as much of the information contained in these two documents can be used to ensure that a survivable EOC is designed and developed. These documents will also be helpful in developing policies and procedures.

The Emergency Operations Plan

As described in the previous chapter, an EOP should describe a jurisdiction's approach to planning—how citizens and property will be protected during a terrorist incident—and the resource identification and management system. An EOP must be effective in turning the concept of operations into an effective emergency response.

1. Highlight or circle each task you perform or would be expected to perform.

    a. Assist in the location and design of an EOC.

    b. Form/convene planning team/committee.

    c. Use a hazard/vulnerability analysis to assist in locating/designing an EOC.

    d. Define functions performed in the EOC.

    e. Determine the number of personnel required to operate the EOC.

    f. Determine space requirements for the EOC.

    g. Determine funding requirements for the EOC.

    h. Assess and evaluate functional layout (i.e., operational efficiency) of the EOC.

    i. Develop contingency plan for interim operations.

2. Assist in the preparation of the EOC for operations (i.e., fully functioning capability).

    a. Assist in determining telecommunications requirements.

    b. Assist in defining life support requirements.

    c. Assist in determining operating equipment/supplies needed.

    d. Ensure that procedures are in place to maintain support systems and equipment.

Figure 11.1   Responsibilities for EOC management and operations.

The Hazard and Vulnerability Analysis

All geographical areas of the United States are vulnerable to a terrorist attack, due to the wide variety of means by which an attack could take place. Identification of these hazards must be accompanied by a determination of the risk each hazard poses.

Using the EOP and Hazard Analysis to Design an EOC

Various terrorism-related hazards will have different effects on the ability to survive the event and to continue to direct, control, manage, and coordinate emergency operations both within a jurisdiction, with other state and local governments, and with the federal government.

The design criteria for an EOC depend on the types of terrorism-related disasters that are likely to affect a jurisdiction. Although some terrorist threats are possible in all settings, areas that are either heavily populated or heavily rural, along with those close to military and utility sites, will have different considerations in terms of designing an EOC. The hazard and vulnerability analysis provides a good basis for determining the worst-case scenario in designing the EOC. The most

Figure 11.2 EOC setup. (Courtesy of FEMA, Savannah Brehmer.)

critical element of the EOC is its ability to survive an emergency without any interruption in the continuity of operation (see Figure 11.2).

Protection needs to be an integral part of planning, building, modifying, or equipping your jurisdiction's EOC. Personnel will need to be protected against a wide range of conditions, requiring a close examination of EOC location, structural integrity, and security procedures.

Developing Policies and Procedures

No matter how established your EOC, if you fail to develop policies and procedures for EOC management, your EOC will not be able to sustain itself during a terrorist incident. Effective EOC management and operation is dependent upon written policies and procedures that are in place prior to a terrorism-related emergency. Without these written guidelines, coordinated and responsive efforts cannot be achieved during an incident.

Each EOC will have its own unique requirements, depending on the jurisdiction and its demographics and area; however, certain standard policies and procedures will apply to any EOC. These standard policies relate to:

■ Communications
■ Life support

Figure 11.3   Emergency phones ready to be installed. (Courtesy of FEMA, Patsy Lynch.)

- Equipment and supplies
- Documents and records retention

Communications Policies and Procedures

Adequate communications are essential to a jurisdiction's ability to direct its emergency response personnel effectively in the aftermath of a terrorist attack, regardless of the nature of that attack. Therefore, there must be plans in place for the effective emergency use of the extensive communications systems. You should make a point to consult with communications experts to ensure that your EOC's communication system is fully functional, compatible with other systems, and able to handle the demand of increased calls during an emergency. One critical lesson learned from the events of September 11, 2001, is that a confusion in communication led to further issues at the site, instead of those communications being used to calmly and effectively monitor and lead the response.

Developing Communication Standard Operating Procedures

EOC communications and standard operating procedures (SOPs) cannot resolve all major problems, especially as the possible impact of certain types of terrorist attacks are still being studied so as to be fully understood from an emergency management perspective, but properly designed procedures can lessen their severity. Some typical SOPs include the following:

- Available communications systems
- Frequencies listing

- Call-up and alerting procedures for communications personnel
- Location and communications supplies and equipment
- Setup procedures for communications equipment
- Tasks and responsibilities for communications personnel
- Procedures for internal face-to-face communication (e.g., relaying messages)

Life Support Policies and Procedures

In some emergency situations, EOC staff may be isolated for an extended period of time. Therefore, it is important to have life support systems to ensure that EOC personnel can sustain for a period of up to two weeks.

Identify Emergency Operations Center Life Support Requirements

No matter where your jurisdiction is located, there are numerous agencies, corporations, and volunteer groups that will be willing to assist you in planning for terrorist attack response operations. Some may even provide goods and services free of charge or at a reduced cost. If you work with vendors that require payment, be sure to establish a memorandum of agreement or some other type of legal agreement or procurement contract to ensure that goods and services will be available when you need them. A lack of organization or expectation of what is to be provided is particularly damaging when trying to deal with the emergency management response required from a terrorist attack.

Tracking Goods and Services Required for the EOC

Once you have identified sources of the goods and services you will require for your EOC, you should maintain a list of applicable names and phone numbers. The following is a list of minimum life support requirements for the EOC.

- Sleeping accommodations—Some type of sleeping arrangements for EOC personnel will need to be made. Two- or three-tier bunks can be used to conserve space. Sleeping bags and portable bedding should be stored in a location near the EOC.
- Food service—Plan on having enough food to sustain EOC personnel for a minimum of fourteen days.
- Water—Arrangements for water should be in place. Faucet water may be undrinkable, especially considering the possible ways in which a terrorist attack could occur or unfold, so have bottled water or water sanitation tablets. Plan for a minimum of ten gallons of water per day, per person. Water must also be available for showers, waste disposal systems, and so forth.

- Sanitary facilities—Facilities such as toilets, showers, laundry, and garbage disposal should be provided but located away from the operations area. Supplies, such as toilet paper, should be readily available to meet demand caused by having more people using the facility.
- Medical supplies—Medical supplies should be limited to those required for dispensary-type operation. First-aid kits and an extra supply of bandages and antiseptics should be priority items, as well as medicines to treat diarrhea, headaches, constipation, and minor respiratory problems.
- Heating, ventilation, and air-conditioning (HVAC)—Facilities should be comfortable at all times. Ensure that procedures are in place to compensate for failed HVAC conditioning systems. Battery-operated backup equipment should be available.

There may be some federal, state, and local standards for life support systems. You should check with your records and documents office for up-to-date information regarding these standards. If you fail to do so, you may not be in compliance. These standards help to ensure that you maintain a minimally acceptable level of performance and operations for life support systems.

Life support systems are an essential component of successful emergency operations. Therefore, it is important that a regular schedule of training and exercising be implemented. Postincident critiques should be developed and fed back into your planning efforts.

Operating Equipment and Supplies Policies and Procedures

There are no set standards for optimum supply requirement to operate the EOC. Requirements will depend on the type of incident, the number of staff that will be working at the EOC, the size of the EOC, and other factors too numerous to list here. If historical data for your community exist, it may be a good idea to review this information as a means to estimate needs. Keep in mind, however, that estimates for fully functioning equipment and sufficiency of supplies should be based on a two-week operational period. Although other terrorist incidents may not be so severe (and that can't safely be assumed), consider the amount of time needed for the response to the September 11, 2001, attacks and be mindful that supplies could be needed for a long period of time.

The more attention given to acquiring appropriate supplies prior to an emergency, the less you will need to rely on a backup plan for obtaining supplies. In selecting equipment, whether for initial purchase or backup plans, consider the following.

- Mobility—Since conditions change during a crisis, the EOC's configuration must be flexible. Bulky, heavy, or cumbersome supplies and equipment should not be used.

- Reliability—Equipment should be durable and reliable. If the equipment has a tendency to break down, then spare parts, instructions for repair, and training for maintenance should be included in the planning effort and—where possible—incorporated into the SOPs. Batteries should be rotated.
- Electrical compatibility—All electrical equipment should be tested on the EOC power supply. All equipment operating standards should be carefully compared to emergency power standards.
- Sustainability—Primary power sources often fail, so you should plan for backup methods to operate equipment. Similarly, you must have backup supplies to replace alternate power sources, such as batteries or generators. The following should be in place:
  - Spare parts inventory for backup lighting, communications, ventilation, and other necessary maintenance.
  - Auxiliary lighting, such as flashlights, batteries, and bulbs.
  - Office materials, including an adequate supply of forms, pencils, paper clips, tape, note pads, and so forth. Computer equipment should not be dependent on outside data banks because they could fail. Keep manual typewriters available in case of an electrical power failure.
  - Recording equipment, such as instant cameras and battery-operated recorders.
  - Specialized equipment that may be required for certain hazards, such as a hazardous materials incident that occurs as the result of a terrorist attack.

Once that initial equipment has been acquired, you should have SOPs in place that address maintenance and acquisition of backup operating equipment and supplies.

Maintenance Contracts

Inspection and maintenance schedules are used to ensure that all equipment is in operating condition, when needed. You don't need an elaborate system to keep track of inspection and maintenance schedules, but it is important to have some type of system in place to monitor compliance with these schedules.

If you have access to a computer, you should use a database tracking system for developing and maintaining the schedules. Any scheduling system you use should contain the following elements:

- Description of equipment, including model number, serial number, and manufacturer
- Names, addresses, and phone numbers for vendors
- Contract number and account information
- Date equipment was purchased or leased
- Date of last inspection or maintenance

- Date of next inspection or maintenance
- Expiration dates of contracts

If maintenance contracts are not in place for critical pieces of equipment, immediate arrangements should be made to secure them. You can purchase the equipment or other types of resources, or you can establish memorandums of understanding to ensure that you can acquire the equipment during an emergency.

Memorandums of Understanding and Mutual Aid Agreements

A memorandum of understanding (MOU) is an agreement between agencies—both internal and external—located within the jurisdiction on cooperative efforts and services that would be provided during an emergency response. The agencies involved usually maintain command of their personnel while providing specific services to the community at large and in conjunction with the normal resources available in the community.

Mutual aid agreements ensure that you have resources and logistical support available to assist you in managing the response to a terrorist incident. They must be written in accordance with existing state and local ordinances. They should include a discussion of free access across boundaries, command of resources and staff, compensation for workers, staff, support provision, and insurance. Some states even have master mutual aid agreements to which your jurisdiction may subscribe. All MOUs and mutual aid agreements should be reviewed on a regular basis to ensure currency and also that the terms are still in effect.

Records and Document Retention

It is virtually impossible to accurately and properly complete the necessary record keeping after disaster or emergency work has been completed and a period of time has elapsed. Therefore, the importance of prompt, efficient record keeping cannot be overemphasized. You must know what records to keep, how to keep them, and have someone familiar enough to start keeping these records at the onset of an emergency situation.

If the situation following a terrorist attack develops into a major disaster declaration, proper documentation will be needed to justify local expenditures for which reimbursement will be requested. Without proper record keeping, the EOC could lose considerable sums of money because claims for reimbursement will also be needed to justify expenditures for which reimbursements will not be requested.

Most EOC records retention and archiving fall into one of the following categories:

- Survivable records and databases needed to conduct emergency operations
- Survivable records needed to reconstitute the government and for recovery
- A continuity-of-government plan, including an approved succession plan

It is essential that the information requirements for disaster response in your state and local community be identified and cataloged. Primary and alternate EOCs should contain the information databases and records necessary to sustain emergency operations, with provisions made for backup of this data.

Although information needs will vary from jurisdiction to jurisdiction, there are general categories of information that each EOC should maintain. For each type of record listed, the relevant information on contact persons, procedures for contact, location, purpose, and other appropriate information should be organized so that it is easily accessible in an emergency.

- Alerting
- Notification
- EOC activation
- Sheltering
- Transportation
- Food and water supply
- Medical assistance
- Debris removal
- Damage assessment
- Disaster assistance
- Public information

The state, in collaboration with local jurisdictions, must have a system for secure storage of vital records necessary to reconstitute the government and to conduct recovery efforts.

Summary

Preparing your jurisdiction's EOC may seem like a daunting task, but it is absolutely necessary to ensure an effective response to any terrorism-related incident for your jurisdiction. It is important to remember that there are resources available to you and your jurisdiction. The Emergency Management Institute in Emmitsburg, Maryland, provides training to enhance U.S. emergency management practices through a nationwide program of instruction.

Chapter 12

Continuity of Operations Planning

Chad Bowers

Contents

Continuity of Operations Planning

Continuity of operations planning (COOP) facilitates the performance of essential functions during all-hazards emergencies or other situations that may disrupt normal operations. Continuity planning is a fundamental responsibility of public institutions and private entities to our nation's citizens. Continuity planning facilitates the performance of essential functions during an emergency situation that disrupts normal operations and the timely resumption of normal operations once the emergency has ended. A strong continuity plan provides the organization with the means to address the numerous issues involved in performing essential functions and services during an emergency. Without detailed and coordinated continuity plans, and effective continuity programs to implement these plans, jurisdictions risk leaving our nation's citizens without vital services in what could be their time of greatest need.

It is the policy of the United States to maintain a comprehensive and effective continuity capability composed of continuity of operations and continuity of government programs to ensure the preservation of our form of government under the Constitution and the continuing performance of government operations and functions under all conditions as outlined in the following:

- National Security Presidential Directive 51 (NSPD-51)
- Homeland Security Presidential Directive 20 (HSPD-20)
- Federal Continuity Directive 1 (FCD1)
- Federal Continuity Directive 2 (FCD2)
- Continuity Guidance Circular 1 (CGC1)

Continuity requirements must be incorporated into the daily operations of all agencies to ensure seamless and immediate continuation of mission essential function (MEF)/primary mission essential function (PMEF) capabilities so that critical government functions and services remain available to the nation's citizens.

Continuity planning is the good business practice of ensuring the execution of essential functions under all circumstances. Continuity includes all activities conducted by jurisdictions to ensure that their essential functions can be performed. This includes plans and procedures that delineate essential functions, specify succession to office and emergency delegation of authority; provide for the safekeeping of vital records and databases; identify alternate operating strategies; provide for continuity communications; and validate these capabilities through test, training, and exercise (TT&E) programs. Today's changing threat environment and the potential for no-notice emergencies, including localized acts of nature, accidents, technological system failures, and military or terrorist attack-related incidents, have increased the need for continuity capabilities and planning across all levels of government and the private sector.

The goal of continuity planning is to reduce the consequence of any disruptive event to a manageable level. The specific objectives of a particular organization's continuity plan may vary, depending on its mission and functions, its capabilities, and its overall continuity strategy. In general, continuity plans are designed to:

- Minimize loss of life, injury, and property damage.
- Mitigate the duration, severity, or pervasiveness of disruptions that do occur.
- Achieve the timely and orderly resumption of essential functions and the return to normal operations.
- Protect essential facilities, equipment, records, and assets.
- Be executable with or without warning.
- Meet the operational requirements of the respective organization. Continuity plans may need to be operational within minutes of activation, depending on the essential function or service, but certainly should be operational no later than twelve hours after activation.
- Meet the sustainment needs of the respective organization. An organization may need to plan for sustained continuity operations for up to thirty days or longer, depending on resources, support relationships, and the respective continuity strategy adopted.
- Ensure the continuous performance of essential functions and operations during an emergency, including those such as pandemic influenza that require additional considerations beyond traditional continuity planning.
- Provide an integrated and coordinated continuity framework that takes into consideration other relevant organizational, governmental, and private sector continuity plans and procedures.

Responsibility for continuity planning resides with the highest level of management of the organization involved. The senior elected official or the administrative head of a state or local organization is ultimately responsible for the continuation of essential services during an emergency and for the related planning. Organizational responsibilities typically include the development of the strategic continuity vision and overarching policy, the appointment of key continuity personnel, and the development of a program budget that provides for adequate facilities, equipment, and training.

Organizational continuity planning cannot be approached in isolation. The effectiveness of one continuity plan is often dependent upon the execution of another organization's continuity plan as many agency functions rely on the availability of resources or functions controlled by another organization. Such interdependencies routinely occur between government and private sector organizations. Likewise, many government continuity plans are dependent upon private sector resources, especially in the area of critical infrastructure and key resources support.

Effective implementation of continuity plans and programs requires the support of senior leaders and decision makers who have the authority to commit the organization and the necessary resources to support the programs. Emergency management officials are often responsible for developing or assisting in the development of continuity plans and programs for their jurisdictions. They are also available to assist in reestablishing essential functions and services during emergencies and disasters.

An organization's resiliency is directly related to the effectiveness of its continuity capability. An organization's continuity capability—its ability to perform its essential functions continuously—rests upon key components and pillars, which are in turn built on the foundation of continuity planning and program management. These pillars are leadership, staff, communications, and facilities. The continuity program staff within an organization should coordinate and oversee the development and implementation of continuity plans and supporting procedures.

Pillars 1 and 2: People/Leadership and Staff

Continuity of leadership is critical to ensure continuity of essential functions. Organizations should provide for a clear line of succession in the absence of existing leadership and the necessary delegations of authority to ensure that succeeding leadership has the legal and other authorities to carry out their duties. Continuity of leadership during crisis, especially in the case of senior positions, is important to reassure the nation and give confidence to its citizens that the principal or appropriate successor is managing the crisis and ensuring the performance of essential functions. Leaders need to set priorities and keep focus.

Leaders and staff should be sufficiently trained to be able to perform their duties in a continuity environment. To ensure that required skill sets are available, personnel should be both cross-trained and vertically trained to be able to perform the functions of their peers and the persons above and below them in an emergency.

Pillar 3: Communications and Technology

The ability to communicate is critical to daily operations and absolutely essential in a crisis. The nation's domestic and international telecommunications resources, including commercial, private, and government-owned services and facilities, are essential to support continuity plans and programs. All organizations should identify the communication requirements needed to perform their essential functions during both routine and continuity conditions. Communication systems and technology should be interoperable, robust, and reliable. Planners should consider the resilience of their systems to operate in disaster scenarios that may include power and other infrastructure problems.

Organizations should use technology to perform essential functions as an intrinsic part of daily operations, utilizing voice, data, and video solutions as appropriate. Communications and business systems, including hardware and software for continuity operations, should mirror those used in day-to-day business to assist continuity leadership and staff in a seamless transition to crisis operations.

Pillar 4: Facilities

Facilities are the locations where essential functions are performed by leadership and staff. Organizations should have adequate, separate locations to ensure execution of their functions. Physical dispersion should allow for easy transfer of function responsibility in the event of a problem in one location.

The Foundation: Continuity Planning and Program Management

Although an organization needs leaders, staff, communications, and facilities to perform its essential functions, it also needs well-thought-out and detailed plans for what to do with those key resources. Planning should include all of the requirements and procedures needed to perform essential functions.

Other key continuity concepts include geographic dispersion, risk management, security, readiness, and preparedness. Geographic dispersion of an organization's normal daily operations can significantly enhance the organization's resilience and reduce the risk of losing the capability to perform essential functions. Geographic dispersion of leadership, data storage, personnel, and other capabilities may be essential to the performance of essential functions following a catastrophic event and will enable operational continuity during an event that requires social distancing (e.g., pandemic influenza and other biological events).

Risk management is the process to identify, control, and minimize the impact of uncertain events. Security is a key element to any continuity program to protect plans, personnel, facilities, and capabilities to prevent adversaries from interfering with continuity plans and operations. To ensure the safety and success of continuity operations, an effective security strategy should address personnel, physical, and information security.

Continuity Program Management Cycle

A standardized continuity program management cycle ensures consistency across all continuity programs and supports the foundation and pillars that comprise the nation's continuity capability. It establishes consistent performance metrics, prioritizes implementation plans, promulgates best practices, and facilitates consistent cross-agency continuity evaluations. Such a cyclic-based model that incorporates planning, training, evaluating, and the implementation of corrective actions gives key leaders and essential personnel the baseline information, awareness, and

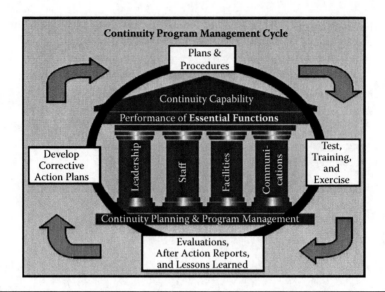

Figure 12.1   Continuity program management cycle.

experience necessary to fulfill their continuity program management responsibilities. The continuity program management cycle consists not only of its programmatic elements, but also should include the plans and procedures that support implementation of the continuity program. These plans and procedures should also be evaluated pre- and postevent, tested or exercised, and assessed during the development of corrective action plans. Objective evaluations and assessments, developed from tests and exercises, provide feedback on continuity planning, procedures, and training. This feedback in turn supports a corrective action process that helps to establish priorities, informs budget decision making, and drives improvements in plans and procedures. This continuity program management cycle, as illustrated in Figure 12.1, should be used by all organizations as they develop and implement their continuity programs.

To support the continuity program management cycle, organizations should develop a continuity multiyear strategy and program management plan that provides for the development, maintenance, and annual review of continuity capabilities, requiring an organization to:

■ Designate and review MEFs and PMEFs, as applicable
■ Define both short-term and long-term goals and objectives for plans and procedures
■ Identify issues, concerns, and potential obstacles to implementing the program, as well as a strategy for addressing these, as appropriate
■ Establish planning, training, and exercise activities, as well as milestones for accomplishing these activities

- Identify the people, infrastructure, communications, transportation, and other resources needed to support the program.
- Forecast and establish budgetary requirements to support the program
- Apply risk management principles to ensure that appropriate operational readiness decisions are based on the probability of an attack or other incident and its consequences
- Incorporate geographic dispersion into the organization's normal daily operations, as appropriate
- Integrate the organization's security strategies that address personal, physical, and information security to protect plans, personnel, facilities, and capabilities, to prevent adversaries from disrupting continuity plans and operations.

Each organization should develop a corrective action program (CAP) to assist in documenting, prioritizing, and resourcing continuity issues identified during TT&E, assessments, and emergency operations.

Essential Functions

All organizations should identify and prioritize their essential functions as the foundation for continuity planning. Essential functions, broadly speaking, are those functions that enable an organization to provide vital services, exercise civil authority, maintain the safety of the general public, and sustain the industrial/economic base during an emergency.

The identification and prioritization of essential functions are a prerequisite for continuity planning, because they establish the planning parameters that drive an organization's efforts in all other planning and preparedness areas. Resources and staff will likely be limited during an event that disrupts or has the potential to disrupt normal activities and that necessitates the activation of continuity plans, preventing the organization from performing all of its normal functions or services. Therefore, a subset of those functions that are determined to be critical activities are defined as the organization's essential functions. These essential functions are then used to identify supporting tasks and resources that should be included in the organization's continuity planning process.

The National Continuity Policy Implementation Plan has established three categories of essential functions: national essential functions (NEFs), primary mission essential functions (PMEFs), and mission essential functions (MEFs). The ultimate goal of continuity in the federal executive branch is the continuation of NEFs. To achieve that goal, the objective for nonfederal entities is to identify their MEFs and PMEFs, as appropriate, and ensure that those functions can be continued throughout, or resumed rapidly after, a disruption of normal activities.

The eight NEFs represent the overarching responsibilities of the federal government to lead and sustain the nation and will be the primary focus of the federal government's leadership during and in the aftermath of an emergency.

PMEFs are MEFs that must be performed in order to support the performance of NEFs before, during, and in the aftermath of an emergency. PMEFs need to be continuous or resumed within twelve hours after an event and maintained for up to thirty days or until normal operations can be resumed.

MEFs are a broader set of essential functions that includes not only an organization's PMEFs, but also all other organization functions that must be continued throughout or resumed rapidly after a disruption of normal activities, but that do not rise to the level of being PMEFs. MEFs are those functions that enable an organization to provide vital services, exercise civil authority, maintain the safety of the public, and sustain the industrial/economic base during disruption of normal operations.

When identifying an organization's essential functions and categorizing them as MEFs or PMEFs, organizations with incident management responsibilities must incorporate these into their continuity planning requirements for performing these functions. Integration of continuity planning with incident management planning and operations includes responsibilities delineated in the National Response Framework (NRF) and is linked to an organization's ability to conduct its essential functions.

In short, MEFs are the responsibilities/tasks an organization is required to complete to be considered "operational." Each individual organization has its own distinct operational responsibilities. Therefore, each individual organization will then have to have its own unique list of MEFs it is required to conduct, and all of these functions serve a distinct purpose in ensuring continuity of government. Following is an example list of MEFs from various organizations:

- Transport inmates to/from court proceedings (Police)
- Book and process incoming offenders (Police)
- Conduct daily audit of accounts payable log (Accounting)
- Detect and suppress urban, rural, and wildland fires (Fire)
- Respond to 9-1-1 calls and vehicular injuries (EMS)
- Issue medications to tuberculosis/HIV patients (Health)
- Inspect and maintain water/wastewater system (Public works)

When listing MEFs, organizations first need to decide the frequency of how often the function must be conducted (daily, weekly, monthly). Next, organizations need to identify how many personnel it requires to complete the function as well as any specialized resources it may require to complete (vehicles, tools, software, etc.). Once an organization has added its MEFs to the list, it then needs to prioritize the functions based on their importance of being completed. Organizations must also keep in mind that there may also be laws, ordinances, and regulations that stipulate

Priority	Function	Division
Tier 1 – Functions to be performed, given a *One Day* disruption. (Highest priority to lowest)		
#1	Record and index Land Records, Maps, Trade Names, Armed Forces Discharges	Land Records
#2	Receive and index Birth, Marriage, Death and Burial records	Vital Statistics
#3	Issue Marriage and Civil Union licenses	Vital Statistics
#4	Collect fees on all transactions, print reports, transmit receipts to Accounting	Accounting
#5	Provide access to Public Records (Land Records, Maps, Trade Names, Minutes, etc.) for attorneys, title searchers, genealogists, and public; provide copies of above as required	Administrative Services
Tier 2 – Functions to be performed, given a *One Day–One Week* disruption. (Highest priority to lowest)		
#1	Scan Land Records, Trade Names, Armed Forces Discharges	Land Records
#2	Send biweekly payroll records to Accounting	Accounting
#3	Register notary certificates, name changes; perform certifications	Vital Statistics
#4	Track vacancies in public office and notify state as required	Administrative Services

Figure 12.2   Examples of essential functions by priority.

functions the organization must conduct. An example of how an organization should create and display a list of its essential functions is provided in Figure 12.2; first by length of disruption and then by priority of carrying out the function.

Human Capital

People are critical to the operations of any organization. Choosing the right people for an organization's staff is vitally important, particularly in a crisis situation. Leaders need to set priorities and keep focus. During a continuity event, emergency employees and other special categories of employees will be activated by an organization to perform assigned response duties. One of these categories is continuity personnel, referred to as the emergency relocation group, relocation team, or similarly named group of designated personnel.

An organization's continuity of operations program, plans, and procedures should incorporate existing organization-specific guidance and direction for human capital management. These can include guidance on pay, leave, work scheduling, benefits, telework, hiring, authorities, and flexibilities. An organization's continuity coordinator (or continuity manager) should work closely with the organization's chief human capital officer or director of human resources to resolve human capital issues related to a continuity event. Human capital issues can be solved using available laws, regulations and guidance, as well as organization implementing instructions.

The planning and preparedness related to leadership, staff, and human capital considerations for a continuity of operations situation encompass the following six activities:

- Organizations should develop and implement a process to identify, document, communicate with, and train continuity personnel.
- Organizations should provide guidance to continuity personnel on individual preparedness measures they should take to ensure a coordinated response to a continuity event.
- Organizations should implement a process to communicate the organization's operating status with all staff.
- Organizations should implement a process to contact and account for all staff in the event of an emergency.
- Organizations should identify a human capital liaison—a continuity coordinator or a continuity manager—to work with the organization's human resources and emergency planning staff when developing the organization's emergency plans.
- Organizations should implement a process to communicate their human capital guidance for emergencies (pay, leave, staffing, and other human resources flexibilities) to managers and make staff aware of that guidance in an effort to help organizations continue essential functions during an emergency.

Continuity Teams and Leadership

The following responsibilities are assigned to the leadership of designated entities. Government organizations play an integral role in determining and supporting the needs of the general public and ensuring the continuation of essential services on a daily basis (e.g., police and fire services, road construction, and public education). These organizations should work with their tribal, local, state, and federal partners and the private sector in developing and coordinating continuity plans. This coordination helps facilitate the resourcing and allocation of resources for the development of continuity plans and the procurement of emergency response equipment, as appropriate.

Elected Officials/Executive Continuity Team

At all jurisdictional levels, elected officials are responsible for ensuring that continuity programs are appropriately resourced, and that responsible and effective continuity leaders and managers are appointed or hired to direct those programs. Elected officials should develop an executive continuity team for the jurisdiction that encompasses all of the departments, divisions, or other offices within the jurisdiction. The elected officials should sign off on the final plans and policies developed by the executive continuity team and each of the participating organizations' continuity of operations plans within the jurisdiction.

Planning Team

The continuity planning team coordinates continuity planning and duties for the entire organization. These duties include:

- Overall continuity coordination for the organization
- Providing guidance and support for the development of the organization's continuity plan
- Establishing designated members who serve on the planning team for their organization or office, and will serve as members of the principal continuity coordinating organization and forum for exchanging ideas and information regarding continuity planning, procedures, and resources for that organization
- Coordinating continuity exercises, documenting postexercise lessons learned, and conducting periodic evaluations of organizational continuity capabilities
- Understanding the role that adjacent jurisdictions and organizations might be expected to play in certain types of emergency conditions and what support those adjacent organizations might provide
- Understanding the limits of their continuity resources and support capabilities
- Anticipating the point at which adjacent organizational or mutual aid resources will be required

When developing continuity teams, it is essential that planners include contact information for each member, as well as an individual description of each member's role for serving on the team. An example layout of a continuity planning team has been provided in Figure 12.3.

Team members are responsible for the following:

- Understanding their continuity roles and responsibilities within their respective organizations
- Knowing and being committed to their duties in a continuity environment

COOP Planning Team		
Name/Title	*Telephone/E-Mail*	*COOP Planning Team Role*
Jane E. Davis Planning Chief	W: 605-555-1234 H: 605-555-1234 C: 605-555-1234 jdavis@Monroe.com	As Planning Chief, will review all changes to the COOP plan prior to submission for final approval.
Jennifer Hughes EMS Dept Manager Emergency Services	W: 354-554-6645 H: 458-658-5995 C: 856-959-6857 P: 785-856-9832 Jen@mail.com	Will serve as the main coordinator to the planning team. Responsibilities include scheduling meetings, notifying team members of meetings.
Ben Wallace Staff Coordinator Ambulance Services	W: 774-662-5553 H: 343-223-2233	Will review and update COOP plan documentation on a quarterly basis.
Chris Matson Operations Manager Emergency Services	W: 774-663-5546 H: 647-663-5524 chris@email.com	Responsible for the training of agency personnel in the actions and responsibilities contained within the plan in preparation of a COOP event.
Adam Sharper Training Coordinator Fire and Rescue	W: 854-857-9669 H: 854-54-2545 Asharper@aol.com	Will document all developments and changes for the COOP.

Figure 12.3 Example of continuity team members, contact information, and team roles.

- Understanding and being willing to perform in continuity situations to ensure an organization can continue its essential functions
- Ensuring that family members are prepared for and taken care of in an emergency situation

Orders of Succession

All organizations are responsible for establishing, promulgating, and maintaining orders of succession to key positions. Simply stated, orders of succession can be summed up by the statement "Who comes next?" It is critical to have a clear line of succession to office established in the event leadership becomes debilitated or incapable of performing its legal and authorized duties, roles, and responsibilities.

Leadership Succession for Health Director			
Position	*Title*	*Agency*	*Name (Position currently held by)*
Primary	Health Director	Monroe County Health Department	Bill Callahan
#1 Alternate	Assistant Director	Monroe County Health Department	Joanne Bowers
#2 Alternate	Operations Chief	Monroe County Health Department	Nancy Rogers
#3 Alternate	Manager of Research	Monroe County Health Department	Lucy Kloss

Figure 12.4 Example list of an order of succession.

The designation as a successor enables that individual to act for and exercise the powers of a principal in the event of that principal's death, incapacity, or resignation. Orders of succession enable an orderly and predefined transition of leadership within the organization. Orders of succession are an essential part of a continuity plan and should reach to a sufficient depth and have sufficient breadth—at least three positions deep and geographically dispersed where feasible—to ensure that essential functions continue during the course of any emergency. An example of how an organization's order of succession should be arranged is provided in Figure 12.4

As a minimum, orders of succession should do the following:

■ Establish an order of succession for leadership. There should be a designated official available to serve as acting head until that official is appointed by appropriate authority, replaced by the permanently appointed official, or otherwise relieved.
  – Geographical dispersion to include, if applicable, regional, field, or satellite leadership in the line of succession, is encouraged and ensures roles and responsibilities can transfer in all contingencies.
  – Where a suitable field structure exists, appropriate personnel located outside of the subject region should be considered in the order of succession.
■ Establish orders of succession for other key leadership positions, including administrators, key managers, and other key mission essential personnel or equivalent positions. Order of succession should also be established for devolution counterparts in these positions.
■ Describe orders of succession by positions or titles, rather than by the names of the individuals holding those offices. To ensure their legal sufficiency,

coordinate the development of orders of succession with the general counsel or other comparable legal authority.

■ Establish the rules and procedures designated officials should follow when facing the issues of succession to office.

■ Include in the succession procedures the conditions under which succession will take place in accordance with applicable laws and procedures; the method of notification; and any temporal, geographical, or organizational limitations to the authorities granted by the orders of succession.

■ Include orders of succession in the vital records and ensure they are available at the continuity facilities or other continuity of operations locations in the event the continuity plan is activated.

■ Revise orders of succession, as necessary, and distribute the revisions promptly as changes occur.

Delegations of Authority

To ensure a rapid response to any emergency requiring the implementation of its continuity plan, an organization should delegate authorities for making policy determinations and other decisions, at the field, satellite, and other organizational levels, as appropriate. It is vital to clearly establish delegations of authority, so that all organization personnel know who has the right to make key decisions during a continuity situation. Generally, a predetermined delegation of authority will take effect when normal channels of direction and control are disrupted and will lapse when those channels are reestablished.

Primary Facilities

When creating a continuity plan, it is important to first identify the locations where an organization operates. This information is used to identify the types of amenities and specific requirements currently in place to support the operations of an organization under normal operations.

Alternate Facilities

As part of the continuity planning process, all organizations should identify continuity facilities; alternate uses for existing facilities; and, as appropriate, virtual office options including telework. Risk assessments should be conducted on these facilities to provide reliable and comprehensive data to inform risk mitigation decisions that will allow nonfederal entities to protect assets, systems, networks, and functions while determining the likely causes and impacts of any disruption. All personnel should be briefed on organization continuity plans that involve using or relocating personnel to continuity facilities, existing facilities, or virtual offices. Continuity personnel should be provided supplemental training and guidance on relocation procedures.

A major section of your plan revolves around the concept of identifying an alternate facility or backup location for each of your primary facilities. Imagine if your workspace or building was unusable due to an event. Your organization would be confronted with many immediate questions, including:

- Where do we relocate our operations; what facility?
- What do we need in place at that alternate facility?
- What items are already at the alternate facility?
- What other items do we need to bring, and in what quantities?

Every organization has different needs and requirements to operate; therefore, they need to think these questions through for themselves. An organization should discuss and decide upon as many of these questions as possible ahead of an event to ensure relocation to a new facility is accomplished as efficiently as possible.

An organization should try to identify at least two alternate facility choices for each its primary facilities. The first choice facility should be a facility close to the primary facility and easily accessible assuming a small-scale disruption that only impacts the single building or work area (fire, pipe burst, mold in the walls, etc.). The second choice facility should be on a regional level and should assume a large-scale event has impacted the surrounding area (tornado, hurricane, earthquake, etc.).

For each alternate facility, it is important to provide specific details about the facility, including the resources that are located at this location versus the resources that would need to be transported to this location to continue operations. Resources to identify might include computers, communication equipment, office furniture, emergency supplies, and any other amenities the organization relies upon to operate. Figure 12.5 is an example of how to identify and list the requirements of an alternate facility, including a detailed list of items that are prepositioned at the alternate facility in addition to the extra items that need to be transported to the facility for operations.

In many instances, organizations have a hard time identifying a specific facility ahead of time that would be available for relocation. The fact is there usually isn't empty working space just waiting around for people to move into. This fact shouldn't deter an organization from moving forward with its planning efforts. In this case, it is recommended to address a facility as "to be determined." With this approach, an organization does not immediately have to be able to identify a specific location to relocate, but can at least begin the discussion about the specific types and number of resources needed for operations. Once a list of resources is developed, the planning team should then have a better understanding of the type and size of location its organization would require for operations.

Name Location (Physical Address)	Resources Required To Perform Mission-Essential Functions
Alternate Facility Monroe County Complex 1220 Greeley Street Monroe, FL 45245 **Facility Manager Name & Telephone** Fran Wills 615-334-9090	**Transported** 12 – Desks – *Furniture* 1 – Generators – *Emergency Equipment* 10 – Desktop/Laptop Computers – *Computer Hardware* 4 – Walkie Talkies – *Communications* 1 – VHF Base Station USMS Frequencies – *Communications* 2 – Satellite Phones – *Communications*
Alternate For Monroe County FD Headquarters – *Primary Facility* 270 Main Street Monroe City, FL 52545	**Prepositioned** 3 – Fire cabinets (Tall/3 Drawer) – *Furniture* 1 – Copy Machine – *Computer Hardware* 1 – Projector/Screen – *Computer Hardware* 16 – Phones – *Communications* 1 – Internet/Intranet Connection – *Communications*

Figure 12.5   Example of an alternate facility and list of required resources.

Vital Records/Vital Resources

Another critical element of a viable continuity plan and program includes the identification, protection, and availability of electronic and hardcopy documents, references, records, information systems, and data management software and equipment (including classified and other sensitive data) needed to support essential functions during a continuity situation. Personnel should have access to and be able to use these records and systems to perform essential functions and to reconstitute back to normal organization operations. Organizations should compile a complete list of vital records and resources used for their day-to-day operations. Vital records and resources used by an organization should then be prepositioned at an alternate facility or stored at a back-up location to ensure performance of essential functions upon COOP activation. Figure 12.6 shows an example of vital records for an organization, including the name and description of each vital record shown in a list in order of priority.

Each organization has different functional responsibilities and business needs. An organization should decide which records are vital to its operations and then should assign responsibility for those records to the appropriate personnel, who may be a combination of continuity personnel, personnel in the chief information officer's department, and records management personnel. An effective vital records program should have the following characteristics.

Vital Records/Resources		
	Record Name	*Data*
#1	Rapid Fire Software	Rapid Fire Software is used for continual training and education of our responders to enhance response times and safety measures.
#2	FARSITE	FARSITE is a fire behavior and growth simulator for use on Windows computers. It is used by Fire Behavior Analysts from the USDA and is taught in the S493 course. FARSITE is designed for use by trained, professional wildland fire planners and managers familiar with fuels, weather, topography, wildfire situations, and the associated concepts and terminology.
#3	Genesis Software	Genesis Software manages the historical calls and prior Responses Database.
#4	SMART System Software	The SMART system is the state-operated database that is used for ...
#5	Emergency Call-Out List	Emergency call-out list consisting of personnel contact information

Figure 12.6   Example of vital records and resources listed by priority.

1. An official vital records program:
   - Identifies and protects those records that specify how an organization will operate in an emergency or disaster
   - Identifies those records necessary to the organization's continuing operations
   - Identifies those records needed to protect the legal and financial rights of the organization and citizens
2. A vital records program should be incorporated into the overall continuity of operations plan, and it needs a clear authority to include:
   - Policies
   - Authorities
   - Procedures
   - The written designation of a vital records manager
3. As soon as possible after continuity of operations activation, but recommended within twelve hours of such activation, continuity personnel at the continuity facility should have access to the appropriate media for accessing vital records, such as:
   - A local area network
   - Electronic versions of vital records
   - Supporting information systems and data
   - Internal and external e-mail and e-mail archives
   - Hard copies of vital records
4. Organizations should strongly consider multiple redundant media for storing their vital records.
5. Organizations should maintain a complete inventory of records along with the locations of and instructions on accessing those records. This inventory should be maintained at a backup/offsite location to ensure continuity if the primary site is damaged, destroyed, or unavailable. Organizations should consider maintaining these inventories at a number of different sites to support continuity operations.
6. Organizations should conduct vital records and database risk assessment to:
   - Identify the risks involved if vital records are retained in their current locations and media, and the difficulty of reconstituting those records if they are destroyed
   - Identify offsite storage locations and requirements
   - Determine if alternative storage media is available
   - Determine requirements to duplicate records and provide alternate storage locations to provide readily available vital records under all conditions
7. Appropriate protections for vital records will include dispersing those records to other organization locations or storing those records offsite. When determining and selecting protection methods, it is important to take into account the special protections needed by the different kinds of storage media. Microforms, paper photographs, and computer disks, tapes, and drives, all

require different methods of protection. Some of these media may also require equipment to facilitate access.

8. At a minimum, vital records should be annually reviewed, rotated, or cycled so that the latest versions will be available.
9. A vital records plan packet should be developed and maintained. The packet should include:
   - A hard copy or electronic list of key organization personnel and disaster staff with up-to-date telephone numbers
   - A vital records inventory with the precise locations of vital records
   - Updates to the vital records
   - Necessary keys or access codes
   - Continuity-facility locations
   - Access requirements and lists of sources of equipment necessary to access the records (this may include hardware and software, microfilm readers, Internet access, and dedicated telephone lines)
   - Lists of records-recovery experts and vendors
   - A copy of the organization's continuity of operations plan

   This packet should be annually reviewed with the date and names of the personnel conducting the review documented in writing to ensure that the information is current. A copy should be securely maintained at the organization's continuity facilities and other locations where it is easily accessible to appropriate personnel when needed.
10. The development of an annual training program for all staff should include periodic briefings to managers about the vital records program and its relationship to the organization's vital records and business needs. Staff training should focus on identifying, inventorying, protecting, storing, accessing, and updating the vital records.
11. There should be an annual review of the vital records program to address new security issues, identify problem areas, update information, and incorporate any additional vital records generated by new organization programs or functions, or by organizational changes to existing programs or functions. The review will provide an opportunity to familiarize staff with all aspects of the vital records program. It is appropriate to conduct a review of the vital records program in conjunction with continuity exercises.
12. There should be annual testing of the capabilities for protecting classified and unclassified vital records and for providing access to them from the continuity facility.

Devolution

Devolution is the capability to transfer statutory authority and responsibility for essential functions from an organization's primary operating staff and facilities to other organization employees and facilities, and to sustain that operational capability for an extended period.

Devolution planning supports overall continuity planning and addresses the full spectrum of threats and all-hazards emergency events that may render an organization's leadership or staff unavailable to support, or incapable of supporting, the execution of the organization's essential functions from either its primary location or its alternate location(s). Organizations should develop a devolution option for continuity, to address how the organization will identify and conduct its essential functions during an increased threat situation or in the aftermath of a catastrophic emergency.

Reconstitution

Reconstitution is the process by which surviving or replacement organization personnel resume normal operations from the original or replacement primary operating facility. Reconstitution embodies the ability of an organization to recover from an event that disrupts normal operations and consolidates the necessary resources so that the organization can resume its operations as a fully functional entity. In some cases, extensive coordination may be necessary to procure a new operating facility if an organization suffers the complete loss of a facility or in the event that collateral damage from a disaster renders a facility structure unsafe for reoccupation.

Testing, Training, and Exercising

A well-defined testing, training, and exercise (TT&E) program is necessary to assist organizations to prepare and validate their organization's continuity capabilities and program to perform essential functions during any emergency. This requires the identification, training, and preparedness of personnel capable of performing their continuity responsibilities and implementing procedures to support the continuation of organization essential functions.

Training provides the skills and familiarizes leadership and staff with the procedures and tasks they should perform in executing continuity plans. Tests and exercises serve to assess and validate all the components of continuity plans, policies, procedures, systems, and facilities used to respond to and recover from an emergency situation and identify issues for subsequent improvement. All organizations should plan, conduct, and document periodic tests, training, and exercises to prepare for all-hazards continuity emergencies and disasters, identify deficiencies, and demonstrate the viability of their continuity plans and programs. Deficiencies, actions to correct them, and a timeline for remedy should be documented within an organization's COOP.

Testing

Testing ensures that equipment and procedures are maintained in a constant state of readiness to support continuity activation and operations. An organization's test program should include:

1. Annual testing (at a minimum) of alert, notification, and activation procedures for continuity personnel, with recommended quarterly testing of such procedures for continuity personnel
2. Annual testing of plans for recovering vital records (both classified and unclassified), critical information systems, services, and data
3. Annual testing of primary and backup infrastructure systems and services (e.g., for power, water, fuel) at continuity facilities.
4. Annual testing and exercising of required physical security capabilities
5. Testing and validating equipment to ensure the internal and external interoperability and viability of communications systems, through quarterly testing of the continuity communications capabilities (e.g., secure and nonsecure voice and data communications)
6. Annual testing of the capabilities required to perform an organization's essential functions, as identified in the BPA
7. A process for formally documenting and reporting tests and their results
8. Conducting annual testing of internal and external interdependencies identified in the organization's continuity plan, with respect to performance of an organization's and other organization's essential functions

Training

Training familiarizes continuity personnel with their procedures, tasks, roles, and responsibilities in executing an organization's essential functions in a continuity environment. An organization's training program should include:

1. Annual continuity awareness briefings (or other means of orientation) for the entire workforce
2. Annual training for personnel (including host or contractor personnel) who are assigned to activate, support, and sustain continuity operations
3. Annual training for the organization's leadership on that organization's essential functions, including training on individual position responsibilities
4. Annual training for all organization personnel who assume the authority and responsibility of the organization's leadership if that leadership is incapacitated or becomes otherwise unavailable during a continuity situation
5. Annual training for all predelegated authorities for making policy determinations and other decisions, at the field, satellite, and other organizational levels, as appropriate

6. Personnel briefings on organization continuity plans that involve using or relocating to continuity facilities, existing facilities, or virtual offices
7. Annual training on the capabilities of communications and information technology (IT) systems to be used during an incident
8. Annual training regarding identification, protection, and ready availability of electronic and hardcopy documents, references, records, information systems, and data management software and equipment (including sensitive data) needed to support essential functions during a continuity situation
9. Annual training on an organization's devolution option for continuity, to address how each organization will identify and conduct its essential functions during an increased threat situation or in the aftermath of a catastrophic emergency
10. Annual training for all reconstitution plans and procedures to resume normal organization operations from the original or replacement primary operating facility

Training should prepare continuity personnel to respond to all emergencies and disasters and ensure performance of the organization's essential functions. These include interdependencies both within and external to the organization. As part of its training program, the organization should document the training conducted, the date of training, those completing the training, and by whom.

Exercises

An organization's continuity exercise program focuses primarily on evaluating capabilities or an element of a capability, such as a plan or policy, in a simulated situation. Organizations should refer to the Homeland Security Exercise and Evaluation Program (HSEEP) for additional exercise and evaluation guidance. An organization's exercise program should include:

1. An annual opportunity for continuity personnel to demonstrate their familiarity with continuity plans and procedures and to demonstrate the organization's capability to continue its essential functions
2. An annual exercise that incorporates the deliberate and preplanned movement of continuity personnel to an alternative facility or other continuity location
3. Communications capabilities and both inter- and intraorganization dependencies
4. An opportunity to demonstrate that backup data and records required to support essential functions at continuity facilities or locations are sufficient, complete, and current
5. An opportunity for continuity personnel to demonstrate their familiarity with the reconstitution procedures to transition from a continuity environment to normal activities when appropriate

6. An opportunity for continuity personnel to demonstrate their familiarity with the devolution procedures to reconstitute from a continuity environment to normal activities when appropriate

7. A comprehensive debriefing after each exercise, which allows participants to identify systemic weakness in plans and procedures, and to recommend revisions to the organization's continuity plan

8. A cycle of events that incorporates evaluations, after-action reports, and lessons learned into the development and implementation of a corrective action program (CAP), to include an improvement plan (IP)

9. Organizational participation: conducting and documenting annual assessments of their continuity TT&E programs, and continuity plans and programs

Each organization should develop a CAP to assist in documenting, prioritizing, and resourcing continuity issues identified during TT&E, assessments, and emergency operations. The purpose of CAP is to accomplish the following:

- Identify continuity deficiencies and other areas requiring improvement and provide responsibilities and a timeline for corrective action
- Identify program and other continuity funding requirements for submission to the organization leadership
- Identify and incorporate efficient acquisition processes, and where appropriate, collect all interorganization requirements into one action
- Identify continuity personnel requirements for an organization's leadership and their supporting human resource offices

Phases and Implementation of a Continuity Plan

A continuity plan is implemented to ensure the continuation or rapid resumption of essential functions during a continuity event. Organizations should develop an executive decision-making process that allows for a review of the emergency situation and a determination of the best course of action based on the organization's readiness posture. An organization's continuity implementation process should include the following four phases: readiness and preparedness, activation and relocation, continuity operations, and reconstitution. The four phases are implemented as illustrated in Figure 12.7.

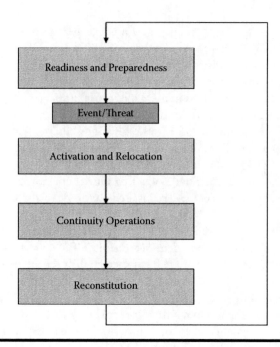

Figure 12.7 Phases and implementation of a continuity plan.

Chapter 13

The Role of Business Continuity Management

Robert Coullahan

Contents

Introduction

Fully 85 percent of the critical assets and key resources comprising our nation's eighteen critical infrastructure sectors are privately owned and operated.* Disaster-driven closures are frequently due to lack of financial reserves for the business disruption disasters bring. The ability to set aside adequate reserves is beyond most businesses. Ninety percent of small businesses hit with a major catastrophe never recover, and only 43 percent of all companies in the same circumstances resume business, ever. Among those that do reopen, only 26 percent are doing business two years later. If their mission-critical data is lost, 50 percent file for bankruptcy immediately.†

Regulations are increasingly compelling publicly held companies to implement effective disaster recovery/business continuity (DR/BC) strategies and plans. Corporate liability insurance rates are increasingly likely to be affected positively when DR/BC solutions are implemented and negatively when they are not. As Brett Williams has succinctly stated: "Crisis, trauma, and disaster are no longer incomprehensible and unimaginable for American businesses. Some industry observers have suggested that organizations are negligent if they do not plan for these situations. No longer can corporate managers and governors claim the defense of the 'unforeseen.'"‡

Organizations owe a duty of care to protect their employees, their communities, and themselves from foreseeable harm. And even where it was not humanly possible to prevent an incident, employees, shareholders, and clients will expect that the management team is prepared to respond quickly and effectively.§

A recent Gartner survey revealed that only 36 percent of organizations surveyed possessed a contingency plan that addressed the complete loss of physical assets at a location. The survey points out that companies are least prepared for the risks that are most likely to impact their organization. In a pre-September 11, pre-Sarbanes–Oxley business environment, some companies would be quite comfortable in the majority of those without a plan for loss of physical assets. After all, there were other seemingly more important issues to focus on like profitability and investor confidence (these issues persist). In a post-September 11, post-Sarbanes–Oxley business environment, this exposure could be very costly.¶

Governments have passed legislation that requires business continuity planning. In addition, insurance underwriters are placing increased pressure on organizations that lack recovery plans by denying them coverage or increasing premiums.

---

* US Department of Homeland Security, National Infrastructure Protection Plan, June 2006.
† Dan Derby, "Early Selling for Business Continuity Planning," *Disaster Recovery Journal* 14, no. 2 (2001): 18.
‡ Brett Williams, "Sarbanes-Oxley: Another Driver for Business Continuity Management," *Disaster Resource Guide*, 2007.
§ Ibid.
¶ Ibid.

Lessons learned in the past decade of natural and human-induced disasters, including terrorist incidents, demonstrates the vital role that the private sector plays in recovery, restoration, and continuity of communities. Enormous resources have been applied to enhance readiness of emergency response agencies and for improvements in public safety. Additional security countermeasures and infrastructure protection solutions have been implemented to protect government services; emergency response capabilities; and responder knowledge, skills, and training. However, the private sector has benefited little from these coordinated efforts, yet it is expected to bounce back rapidly from the consequences of attacks or natural disasters.

Business continuity management has evolved over the past four decades to embrace a more comprehensive preparedness and protection model. What began with emphasis on records management, data recovery, and information technology systems continuity has matured to address all aspects of a business's processes. The convergence of crisis management, business continuity, and infrastructure protection best practices are shaping the emerging domain of "enterprise resilience."

This chapter reviews the ten professional practices of business continuity management and, within the context of an enterprise resilience program, outlines the mission, roles and responsibilities, equipment and technologies that support private sector emergency command center (ECC) operations. The relationship between the ECC and the public safety emergency operations center (EOC) is examined. The importance of private sector owners and operators managing crises with a working knowledge of the Incident Command System (ICS) and the National Incident Management System (NIMS) is also reviewed. Finally, the role of state and local data fusion centers and their liaison with infrastructure owners and operators is examined in the context of advanced data and information sharing tools.

Context and Challenge: Critical Infrastructure Protection

Homeland Security Presidential Directive 7 (HSPD-7): Critical Infrastructure Identification, Prioritization, and Protection, released on December 17, 2003, outlined the requirements for protecting the nation's critical infrastructure.* Attacks on critical infrastructure (CI) could disrupt the direct functioning of key business and government activities, facilities, and systems, as well as have cascading effects throughout the nation's economy and society. The National Infrastructure Protection Plan (NIPP) updated in 2009 describes eighteen critical infrastructure sectors.†

The Commercial Facilities Sector includes a wide range of business, commercial, residential, and recreational facilities where large numbers of people congregate.

---

* Executive Office of the President, Homeland Security Presidential Directive 7 (HSPD-7), Critical Infrastructure Identification, Prioritization, and Protection, December 2003.
† US Department of Homeland Security, National Infrastructure Protection Plan, June 2009.

Commercial facilities operate on the principle of open public access, environments in which the general public may move freely without the deterrent of highly visible security barriers. The Commercial Facilities Sector is diverse in both scope and function, and within the various business lines, each owner/operator has distinct assets, business environments, operational processes, and risk management approaches. The Commercial Facilities Sector comprises eight subsectors*:

■ Entertainment and Media
■ Lodging
■ Outdoor Events
■ Public Assembly
■ Real Estate
■ Resorts
■ Retail
■ Sports Leagues

National Infrastructure Protection Plan (NIPP)

The overarching goal of the NIPP is to

> Build a safer, more secure, and more resilient America by enhancing protection of the Nation's CI/KR to prevent, deter, neutralize, or mitigate the effects of deliberate efforts by terrorists to destroy, incapacitate, or exploit them; and to strengthen national preparedness, timely response, and rapid recovery in the event of an attack, natural disaster, or other emergency.†

The NIPP provides the unifying structure for the integration of existing and future CI/KR protection efforts into a single national program to achieve this goal. The NIPP framework enables the prioritization of protection initiatives and investments across sectors to ensure that government and private sector resources are applied where they offer the most benefit for mitigating risk by lessening vulnerabilities, deterring threats, and minimizing the consequences of terrorist attacks and other manmade and natural disasters.

Protection includes actions to mitigate the overall risk to CI/KR assets, systems, networks, functions, or their interconnecting links resulting from exposure, injury, destruction, incapacitation, or exploitation. In the context of the NIPP, this

---

* US Department of Homeland Security, Summary of National Infrastructure Protection Program Sector-Specific Plans, May 2007.
† US Department of Homeland Security, "Executive Summary," National Infrastructure Protection Plan, June 2009, 1.

includes actions to deter the threat, mitigate vulnerabilities, or minimize consequences associated with a terrorist attack or other incident. Protection can include a wide range of activities, such as hardening facilities, building resiliency and redundancy, incorporating hazard resistance into initial facility design, initiating active or passive countermeasures, installing security systems, promoting workforce surety programs, and implementing cyber security measures, among others.

Achieving the NIPP goal requires actions to address a series of objectives that include:

- Understanding and *sharing information* about terrorist threats and other hazards
- Building *security partnerships* to share information and implement CI/KR protection programs
- Implementing a long-term *risk management program*
- Maximizing *efficient use of resources* for CI/KR protection

These objectives require a collaborative partnership between and among a diverse set of security partners, including the federal government; state, territorial, local, and tribal governments; the private sector; international entities; and nongovernmental organizations. The NIPP provides the framework that defines the processes and mechanisms that these security partners should use to develop and implement the national program to protect CI/KR across all sectors over the long term.

Sector-Specific Plans

The NIPP and supporting sector-specific plans (SSPs) provide the coordinated approach that will be used to establish national priorities, goals, and requirements for CI/KR protection. This coordinated approach will allow federal funding and resources to be applied in the most effective manner to reduce vulnerability, deter threats, and minimize the consequences of attacks and other incidents.

The NIPP establishes overarching concepts relevant to all CI/KR sectors identified in HSPD-7. The SSPs provide details on how the CI/KR mission will be coordinated, developed, and implemented within the seventeen CI/KR sectors. These plans will continue to evolve as threats change and protective programs are implemented. Because each sector has unique issues and concerns, preferred approaches to protection vary within and across CI/KR sectors. SSPs are tailored to address the unique characteristics and risk landscapes of each sector while also providing consistency for protective programs, public and private protection investments, and resources.*

---

* US Department of Homeland Security, Summary of National Infrastructure Protection Program Sector-Specific Plans, Washington, DC, 11 May 2007.

SSPs serve to:

■ Define sector security partners, authorities, regulatory bases, roles and responsibilities, and interdependencies
■ Establish or institutionalize already existing procedures for sector interaction, information sharing, coordination, and partnership
■ Establish the goals and objectives, developed collaboratively with security partners, required to achieve the desired protective posture for the sector
■ Identify international considerations
■ Identify the sector-specific approach or methodology that sector-specific agencies (SSAs), in coordination with the Department of Homeland Security (DHS) and other security partners, will use to implement risk management framework activities consistent with the NIPP

The SSPs are sector-based documents in which the SSA coordinates with the sector's Government Coordinating Council (GCC), Sector Coordinating Council (SCC), and other security partners to develop a plan that highlights the sector-specific approach to CI/KR protection. In future years, the SSAs, in coordination with the GCCs and SCCs, will establish the mechanism to coordinate SSP revision and update in accordance with the process established in the NIPP. DHS, as the overall national coordinator for CI/KR protection, regularly monitors NIPP and SSP implementation actions, and tracks progress and success toward achieving the NIPP goal and objectives.

The Commercial Facilities Sector-Specific Plan (CF-SSP) establishes a relationship between the government and the private sector to foster the cooperation necessary to improve the protection of the sector from a natural or manmade disaster. The CF-SSP sets a path forward for the sector to collectively identify and prioritize its assets, assess risk, implement protective programs, and measure the effectiveness of its protective programs.

The Commercial Facilities Sector security goals are to*:

■ Enable trusted and protected information sharing between public and private security partners at all levels of government
■ Ensure that the public sector security partners disseminate timely, accurate, and threat-specific information and analysis throughout the sector
■ Preserve the "open access" business model of most commercial facilities while enhancing overall security
■ Maintain a high level of public confidence in the security of the sector
■ Provide security that meets the needs of the public, tenants, guests, and employees while ensuring the continued economic vitality of the owners, investors, lenders, and insurers

---

* Ibid.

- Have systems in place (e.g., emergency preparedness, training, crisis response, and business continuity plans) to ensure a timely response to and recovery from natural or manmade incidents
- Institute a robust sectorwide research and development program to identify and provide independent third-party assessments of methods and tools for sector protective program activities

Role of the Private Sector in Contemporary Emergency Operations

Emergency management is defined as the process of preparing for, mitigating, responding to and recovering from an emergency. Emergency management is a dynamic process. Planning, though critical, is not the only component. Training, conducting drills, testing equipment and coordinating activities with the community are other important functions.*

Nowhere in this definition of emergency management are the mission, roles, or responsibilities limited to those of the public safety sector or government agencies. Comprehensive emergency management must include a close and effective public–private partnership to assure continuity of government and continuity of communities.

The Federal Emergency Management Agency (FEMA) has published the Emergency Management Guide for Business and Industry.† It offers a step-by-step approach to emergency planning and recovery for companies of all sizes. The guide is organized as follows:

- Section 1, four Steps in the Planning Process—This section addresses how to form a planning team, how to conduct a vulnerability analysis, how to develop a plan, and how to implement the plan. The information can be applied to virtually any type of business or industry.
- Section 2, Emergency Management Considerations—This section addresses how to build such emergency management capabilities as life safety, property protection, communications, and community outreach.
- Section 3, Hazard-Specific Information—This section provides technical information about specific hazards that a facility may face.
- Section 4, Information Sources—This section identifies where to turn for additional information.

---

* Federal Emergency Management Agency (FEMA), 2007. www.fema.gov
† Federal Emergency Management Agency (FEMA), "Emergency Management Guide for Business and Industry: A Step-by-Step Approach to Emergency Planning, Response and Recovery for Companies of All Sizes," FEMA 141, October 1993.

In community-wide emergencies, business and industry are often needed to assist the community with:

■ Personnel
■ Equipment
■ Shelter
■ Training
■ Storage
■ Feeding facilities
■ EOC facilities
■ Food, clothing, building materials
■ Funding
■ Transportation

Although there is no way to predict what demands will be placed on a company's resources, responsible business leaders will give thought to how the community's needs might influence their corporate responsibilities in an emergency. Indeed they will consider the opportunities for community service before an emergency occurs. Responsible corporate management includes comprehensive business continuity management.

Professional Practices of Business Continuity Management

Business continuity management is defined as a holistic management process that identifies potential impacts that threaten an organization and provides a framework for building resilience with the capability for an effective response that safeguards the interests of its key stakeholders, reputation, and value-creating activities. Its primary objective is to allow the executive of an organization to continue to manage its business under adverse conditions, by the introduction of appropriate resilience strategies, recovery objectives, business continuity and crisis management plans in collaboration with, or as a key component of, an integrated risk management initiative.

The professional development of the business continuity practice continues to mature. Both the Disaster Recovery Institute (DRI) International and the Business Continuity Institute (BCI) figure prominently in the historical evolution of business continuity and disaster recovery professional development and board certification.

The DRI International, a not-for-profit professional association based in Washington, DC, has structured a foundational training and certification program for the business continuity professional community of practice. DRI International was founded in 1988 as the Disaster Recovery Institute in order to develop a base of knowledge in contingency planning and the management of risk, a rapidly growing profession. Today DRI International administers the industry's premier educational and certification programs for those engaged in the practice of business continuity planning and management. More than 3,500 individuals throughout the world maintain professional certification through DRI International.

Individual certification and establishing a common body of knowledge serve to enhance the profession of business continuity management and provide the private sector with consistent, standards-based program designs for business continuity and protection.

As a key institutional actor in business continuity management, DRI International's goals are to:

- Promote a base of common knowledge for the business continuity planning/ disaster recovery industry through education, assistance, and publication of the standard resource database
- Certify qualified individuals in the discipline
- Promote the credibility and professionalism of certified individuals

DRI International offers a range of certification options including:

- Associate Business Continuity Professional
- Certified Functional Continuity Professional
- Certified Business Continuity Professional
- Master Business Continuity Professional

DRI International published the original *Common Body of Knowledge* in September of 1993, after more than two years of intensive work by the Certification Commission. Developed by recognized experts in business continuity/disaster recovery planning and education, the document provided a standard resource base outlining the experience required by professionals in the field. Each subject area in the current "Professional Practices" contains a description, the professional's role within the area, and an outline of recommended knowledge within the subject area.

The ten subject areas listed next cover the competencies required by a professional practitioner in order to deliver effective business continuity management. These professional practices are not presented in any particular order of importance or sequence.

1. Project initiation and management—Establish the need for a business continuity management process or function, including resilience strategies, recovery objectives, business continuity and crisis management plans; also including obtaining management support and organizing and managing the

formulation of the function or process either in collaboration with or as a key component of an integrated risk management initiative.

2. Business impact analysis—Identify the impacts resulting from disruptions and disaster scenarios that can affect the organization and techniques that can be used to quantify and qualify such impacts. Identify time-critical functions, their recovery priorities, and interdependencies so that recovery time objectives can be set.

3. Risk evaluation and control—Determine the events and external surroundings that can adversely affect the organization and its resources (facilities, technologies, etc.) with disruption as well as disaster, the damage such events can cause, and the controls needed to prevent or minimize the effects of potential loss. Provide cost-benefit analysis to justify investment in controls to mitigate risks.

4. Developing business continuity management strategies—Determine and guide the selection of business operating strategies for continuation of business within the recovery point objective and recovery time objective, while maintaining critical functions.

5. Emergency response and operations—Develop and implement procedures for response and stabilizing the situation following an incident or event, including establishing and managing an emergency operations center to be used as a command center during the emergency.

6. Developing and implementing business continuity and crisis management plans—Design, develop, and implement business continuity and crisis management plans that provide continuity within the recovery time and recovery point objectives.

7. Awareness and training programs—Prepare a program to create and maintain corporate awareness and enhance the skills required to develop and implement the business continuity management program or process and its supporting activities.

8. Maintaining and exercising business continuity and crisis managements plans—Preplan and coordinate plan exercises, and evaluate and document plan exercise results. Develop processes to maintain the currency of continuity capabilities and the plan document in accordance with the organization's strategic direction. Verify that the plan will prove effective by comparison with a suitable standard, and report results in a clear and concise manner.

9. Crisis communications—Develop, coordinate, evaluate, and exercise plans to communicate with internal stakeholders (employees, corporate management, etc.), external stakeholders (customers, shareholders, vendors, suppliers, etc.), and the media (print, radio, television, Internet, etc.).

10. Coordination with public authorities—Establish applicable procedures and policies for coordinating continuity and restoration activities with external agencies (local, state, national, emergency responders, defense, etc.) while ensuring compliance with applicable statutes or regulations.

These standards are published by the Disaster Recovery Institute International of the USA and the Business Continuity Institute of the United Kingdom, and are used by both bodies in their certification programs.* This international industry standard for the ten professional practices undergoes a rigorous update process annually. Changes are solicited from professionals worldwide; then those submissions are reviewed by a panel of experts for relevance and accuracy. Resulting changes are incorporated regularly to keep the standard up to date.

Enterprise Resilience

Enterprise resilience is a comprehensive program of readiness activities that imbues the best practices of the business continuity management professional practices, combining the planning and management strategies with an operational understanding of the ICS, NIMS, and crisis communications. It is supported with contemporary knowledge, skills, and technical expertise in emergency command center operations and technologies. It is designed to enhance business survivability, accelerate resumption and restoration, and optimize the role of the private sector in support of community readiness and resilience.

Agility, adaptation, and rapid response are the hallmarks of those organizations that achieve success in developing an enduring, diverse portfolio despite a turbulent and stressful environment. In applying the best practices of business continuity management through a structured program built on the ten professional practices, and including a training and exercise component to assure understanding of the ICS and NIMS management protocols, an enterprise can reduce its risk of business interruption and can restore operations and recover full service readiness on a shorter timeline. A valuable asset in the implementation of an enterprise resilience program is found in the emergency command center, or ECC.

Emergency Command Center: Mission, Roles, and Responsibilities

Leadership begins at the top, but empowerment for an adequate response in the protection of the business must exist at all levels of management. Should the private sector be concerned with possible liability claims, if a business fails to plan and train key personnel within the organization prior to an emergency, crisis, or disaster?

Every business should consider how its leadership will communicate, direct, and control the personnel and resources of the enterprise during both the response

---

* Business Continuity Institute, Certification Standards, http://www.thebci.org/certification-standards.htm, Caversham, United Kingdom, 2003–2007.

and recovery phases of an emergency that impacts operations, life safety, or facilities.

What types of command centers are appropriate? All command centers must have clearly defined functions, duties, responsibilities, policies, and plans based on the interactions from internal and external resources, interdependencies, and must function independently.

- When should the command centers be activated?
- Who has authority to activate these centers?
- What are the criteria for activation?
- Who should be notified and how?
- Does the company have an escalation plan in place before command center activation?

Regardless of the number of command centers, leadership must consider:

- Chain of command
- Personnel assignments
- Communications channels and number of units
- Roles, responsibilities, and duties
- Equipment, resources, and so forth
- What ifs

There are typically two command centers in the public sector. These we intimately know as the incident command post (ICP) and the emergency operations center (EOC). Lessons learned from recent natural and technological disasters suggest that up to three distinct command centers may be essential in the private sector response and recovery phases. These three private sector command centers may include:

- Forward command center (FCC)
- Emergency command center (ECC)
- Business continuity center (BCC)

When to activate each of the three command centers is always a key question. Executive leadership must consider many variables, including personnel, advance notice, location, setup time, and nature of event. Predetermined scenarios can help define the thresholds and conditions under which each of the centers is activated. These event-activated decisions will typically consider the event or incident size or scope; its impact on properties and facilities; its impact on community (external event); its severity and projected duration; and the availability of advance notification such as weather forecasts or fire situation reports.

Let's look at the primary mission, roles, and responsibilities of each of the three private sector command centers.

Forward Command Center

- Safe Location—Within a safe distance from the emergency. Distribution and reporting center closest to the emergency.
- Serves as a Mobile Command Center or Temporary Command Post—FCC.
- Primary function: Preservation of life. Do what it takes to rescue people.
- Secondary function: Damage Limitation.
- Keeps the emergency from turning into a crisis or disaster.
- First management-level command center that makes judgments/assessments on escalation of response, such as "activate the ECC."

The FCC responds to the front lines, but what if there is no front line? Where is the front line in a norovirus outbreak? Where is it in a large area coverage biological event, in an earthquake, tornado, or property-wide power outage? FCCs are event based, thus may or may not exist; however, they still need to be funded, trained, and key personnel identified for effectiveness.

The FCC must be scalable; staff requirements vary according to the size and nature of the emergency

It more often than not will not have a predetermined, fixed location. By circumstances of the event location the FCC will exploit available desks, hard-wired communications systems, cameras, and other resources to aid in incident size-up and response operations. An important consideration for the business is the dangers presented in terms of threat to responders' lives when untrained volunteers, inadequate equipment, and improper verbal commands are at work.

Emergency Command Center

Primary roles of the emergency command center (ECC) are:

1. Preservation of life
2. Damage limitation
3. Restoration of operations

The ECC should follow the policies and procedures in the emergency operations manual/emergency response plan (EOM/ERP). Response must consider NIMS/ICS management protocols with public safety partners and have a working understanding of specialized response teams in the medical, public health, and hazardous materials disciplines. An understanding of the escalation of NIMS procedures through unified command, area commands, multi-agency command, and the National Response Framework (NRF) are essential capabilities requisite to function effectively within the contemporary emergency management milieu.

The location of the ECC is typically in the basement or an expansion of an existing operation center, that is, facilities, security or surveillance room, or conference room. Key features of an ECC are that it is physically protected from all hazards; fully equipped on all communications or technology aspects; provided with sufficient commercial power, including uninterruptible power system (UPS) and portable backup power; affords ample space for the ECC team; and easily accessible.

- Primary function—Preservation of life
- Secondary functions—Damage limitation and restoration of business operations
- Response should consider NIPP, NIMS, CDC (Centers for Disease Control and Prevention), NRP (National Response Plan), FEMA, ICS, and local agency responding capabilities and planned integration
- External response may be "immediate," "shortly," "you're next on the list," "sometime soon," "in a few days," or even longer depending upon scale (extent) and number of properties affected within the community

Key ECC staff are positioned when the ECC is activated, regardless of the type and extent of an event.

Working groups are positioned depending upon the specific event or incident type; that is, if no evacuation is needed, then no emergency evacuation assembly area for guests or employees is activated. A working group may or may not exist depending upon the event. A business that is prepared knows which working groups report to which ECC staff member under what circumstances or conditions. Working groups generally consist of midlevel managers who have authority and experience in dealing with operational challenges. They maintain a single focus and location, and assure attention to detail and responsiveness in time and task completion.

The ECC must be designed and managed with an understanding of the key impacts on services; operations; and, most important, on the enterprise's employees, guests/visitors, customers/clients (including physical, psychological, and other dimensions). When an ECC functional design considers both operational staffing and working group deployments that focus upon preservation of life, damage limitation, and restoration of operations, it has taken a positive step toward fulfilling the mission of business continuity.

A successful ECC is based upon comprehensive advance planning and adequate resource allocation in support of these key components:

- Location (low-risk, secure, protected, and accessible facility space)
- Policies, planning, and procedures (business continuity plan and emergency response plan)
- Operability (ease of use, familiarity through training and exercises)
- Duties/responsibilities (known roles and responsibilities)
- Functionality (knowing the six key functions for each role)
- Personnel (experienced managers)
- Training (training and exercise programs that stress and test functions, duties)
- Equipment (essential but not complex tools)

Business Continuity Center

Location of the BCC could be in an executive conference room or boardroom. It is where the crisis management team as defined in the business continuity plan (BCP) should convene to oversee decision making, and direction and control of business resumption and continuity of operations objectives.

It is the senior leadership access point for the secondary function—adaptation, flexibility, and innovative actions—available to overcome any shortcoming or missing link in the BCP. In consonance with the best practices of business continuity management, prior board of directors' approval of actions and authorities delineated within the BCP is an essential planning action.

Several key questions that should be answered about the BCP well before any exercise or real-world event activates the BCC are:

- Do we have a current, management-approved BCP?
- Does the BCP list responses under an "all-hazards" hazard and threat identification?
- Does the BCP help us identify and reduce risks?
- Does the BCP adequately analyze impact on (current) business operations?
- Does the BCP consider interdependencies both internal and external and within the surrounding communities?
- Have all of our crisis management teams and executive leadership personnel seen the BCP?

Emergency Command Center: Equipment and Technologies

The ECC should be built based on proven design criteria drawn from lessons learned in the engineering and operations of EOCs and command centers that have successfully supported emergency and crisis managers over the decades. The FEMA promulgated EOC Assessment Checklist is divided into six sections and is directly applicable to general ECC design criteria*:

- Facility Features—Examines the physical features of the ECC
- Survivability—Examines the ability of the ECC to sustain the effects of a hazard event and continue operations after the event has occurred
- Security—Examines the protection of the ECC facility, its occupants, and communications equipment and systems from relevant hazards
- Sustainability—Examines the ability of the ECC to operate for extended periods of time without interruption

---

* FEMA, EOC Assessment Checklist, Washington, DC.

■ Interoperability—Examines the extent to which the ECC shares common principles of operations and information exchange with all business partners and local, state, and federal authorities
■ Flexibility—Examines the ability of the ECC to adapt the scale and pace of its operations to the demands of specific hazard events

Preparation through an effective organizational design of an ECC that is fully equipped to meet the challenges experienced through an all-hazards event will speak volumes in the minds of the employees, employee families, bond holders, stock holders, your clientele, and the community at large.

The private sector ECC should be equipped with essential technologies and software applications to enable implementation of the BCP and ERP plans, support crisis communications requirements of the executive team, and fully document actions, direction, reports, and results throughout the response and recovery phases. An audit trail of decisions and actions taken is vital in today's litigious environment, and it is essential for insurance claims and any disaster assistance claims that may be filed.

The ECC design should include the following types/classes of equipment, applications and supplies:

■ Communications equipment
  - Telephone (assess how many lines will be needed; no audible ringer; prefer flashing light end instruments)
  - Workstations with Internet access for e-mail services
  - Battery-operated radio for commercial broadcast service
  - Wireless, cell phone; consider backup satellite phones
  - Radio base station (for security force; handhelds for engineering/maintenance staff)
  - Additional handheld UHF/VHF radios and satellite pagers for hand receipt checkout
  - Facsimile machine(s)
  - Audio cassette player
  - VHS player
  - DVD player
  - Copy machine
  - Overhead projector with screen
■ Computer-based tools
  - Geographic information system (GIS) applications
  - Consequence assessment tool (for impact assessments)
  - Mobile data system
  - Incident Management System for situation reporting and resource management.

- Auxiliary software applications for tracking response, recovery, tasks, operations, checklists and personnel
- Document scanner(s)
■ Surveillance and monitoring systems
    - Closed-circuit television (CCTV) cameras
    - Surveillance monitors
    - Video recorders with playback and time/date stamp features
    - Audio recording system
■ Protective equipment
    - Respiratory protection for all personnel (N95 respiratory mask or equivalent*)
    - Protective clothing (e.g., Tyvek® disposable suits)
    - Hazardous Waste Operations and Emergency Response standards
    - First aid kits
    - American Red Cross CPR (cardiopulmonary resuscitation) and AED (automated external defibrillator) trained/certified staff
■ Supporting infrastructure
    - Portable diesel generator(s)
    - Battery backup UPS
    - Gas-powered portable generators
    - Emergency lighting systems
    - Clocks (battery powered)
    - Dry-marker whiteboards
    - Calculators
■ Reference materials
    - Facility drawings, blueprints, computer-aided design (CAD) files
    - Material safety data sheets (MSDS)
    - Hazardous materials location data
    - Emergency evacuation routes/maps/floor plans
    - Emergency response plan
    - Business continuity plan
    - Continuity of operations plan (COOP)
    - Job aids and checklists

---

* The US Department of Labor recommends air-purifying respirators (e.g., N95, N99, or N100) as part of a comprehensive respiratory protection program for workers directly involved with avian influenza–infected birds or patients. N95 respirators have two advantages over simple cloth or surgical masks; they are >95% efficient at filtering 0.3 μm particles (smaller than the 5 μm size of large droplets—created during talking, coughing, and sneezing—which usually transmit influenza) and are fit tested to ensure that infectious droplets and particles do not leak around the mask. Even if N95 filtration is unnecessary for avian influenza, N95 fit offers advantages over a loose-fitting surgical mask by eliminating leakage around the mask. http://www.cdc.gov/ncidod/EID/vol12no06/05-1468.htm.

- Chemical, biological, radiological, nuclear, and high-yield explosives (CBRNE) agent guidance documents (Select agents and toxins list; radiological exposure information, etc.)
- Poison control center contact information and technical data
- Contact database (current data including all public safety EOC contacts, government contacts at local, state, and federal levels; key vendors and customers; and all employees)
■ Expendable supplies
  - Potable water (bottled)
  - Meals ready to eat (MREs)
  - Over-the-counter medications/supplies (aspirin, antacids, antiseptics, bandages)
  - Copier and printer paper
  - Copier and printer ink cartridges and toner
  - Batteries for all portable equipment
  - Flashlights
  - Cleaning supplies and equipment (personnel may become ill)
  - Trash bags
  - Cots, blankets, pillows
  - Personal hygiene kits (toothpaste, soap, deodorant, etc.)
  - Towels, paper towels, tissues, tissue paper

Although these key resources are essential for effective operation of the ECC, the very real potential of being unable to use the primary ECC must be considered in your planning. Prudent planning and lessons learned suggest that a second location should be selected for the alternate ECC. This secondary venue should support the same capabilities and equipment as the primary location. Contamination or destruction of the primary ECC facility, the loss of power, water, or communications services may direct the ECC team to an alternate location. Therefore, it is essential that the enterprise have a shelf-ready continuity of operations plan (COOP) that can be invoked on demand.

COOP is defined as the activities of individual departments and their subcomponents to ensure that their essential functions are performed. This includes plans and procedures that delineate essential functions; specify succession to office and the emergency delegation of authority; provide for the safekeeping of vital records and databases; identify alternate operating facilities; provide for interoperable communications; and validate the capability through tests, training, and exercises. All Federal agencies, regardless of location, should have a viable COOP capability to ensure continued performance of essential functions from alternate operating sites during any emergency or situation that may disrupt normal operations.

COOP planning is simply a "good business practice"—part of the fundamental mission of responsible enterprises. Today's changing threat environment and the potential for no-notice emergencies, including localized acts of nature, accidents, technological emergencies, and military or terrorist attack-related incidents, have increased the need for COOP capabilities and plans that enable organizations to continue their essential functions across a broad spectrum of emergencies.

Relationship with the Public Safety EOC

The *Disaster Recovery Journal* (DRJ), in cooperation with DRI International, has been developing a generally accepted practices (GAP) guide. Drafted in 2007, the document is based upon the ten professional practices. It is designed to be recognized as a leading source of sound, generally accepted practices by providing a depository of knowledge and recommendations offered by skilled business continuity professionals. Best practices have been compiled from submittals by experienced business continuity professionals from the public and private sectors, as well as user groups and related organizations, in regard to the industry standard professional practices. The Best Practice Committee will review the submittals quarterly for approval.

The generally accepted business continuity practices subject areas align with the ten professional practices of business continuity management*:

1. Project Initiation and Management
2. Risk Evaluation and Control
3. Business Impact Analysis
4. Developing Business Continuity Strategies
5. Emergency Response and Operations
6. Developing and Implementing Business Continuity Plans
7. Awareness and Training Programs
8. Maintaining and Exercising Business Continuity Plans
9. Public Relations and Crisis Coordination
10. Coordination with Public Authorities

The professional practices tell you *what* you need to do and the generally accepted business continuity practices will tell you *how* to do it.

Within the GAP document resides valuable guidance on coordination with the public authorities. The document provides a matrix of procedures and policies for coordinating response, continuity, and restoration activities with local authorities while ensuring compliance with applicable statutes or regulations. This includes identifying applicable laws and regulations governing emergency response; identifying agencies supporting disaster recovery and business continuity; and developing plans to meet statutory requirements (http://www.drj.com/GAP/gap.pdf).

---

* Disaster Recovery Journal and DRI International, Generally Accepted Practices for Business Continuity Practitioners, August 2007, p. 3, http://www.drj.com/GAP/gap.pdf.

Emerging Capabilities: State and Local Data Fusion Centers

Effective prevention efforts depend on the ability of all levels/sectors of government and private industry to collect, analyze, disseminate, and use homeland security-related intelligence. This capacity, known as the "fusion process," serves as the foundation for all state and urban area fusion centers. Accordingly, the establishment of a network of fusion centers to facilitate effective nationwide homeland security information sharing was a top prevention priority in fiscal year (FY) 2007.

Funds from the FY 2007 Homeland Security Grant Program (HSGP), including the State Homeland Security Program (SHSP), the Urban Area Security Initiative (UASI), and the Law Enforcement Terrorism Prevention Program (LETPP) utilized to establish/enhance state and local fusion centers must support the following*:

- Development of a statewide fusion process that corresponds with the Global Justice/Homeland Security Advisory Council (HSAC) Fusion Center Guidelines
- Achievement of baseline levels of capability as defined by the Fusion Capability Planning Tool

The Fusion Center Planning Tool was developed based on the Global Justice/ HSAC Fusion Center Guidelines and provides a streamlined framework of critical fusion process capabilities—the achievement of which will result in each fusion center meeting baseline operational standards. Grantees are encouraged to use the Fusion Capability Planning Tool to determine and prioritize areas of improvement, develop strategies to overcome shortfalls, and prioritize the expenditure of funds to address identified areas of improvement.

As noted, the establishment of a baseline capability level within all state and urban area fusion centers is the primary emphasis of FY 2007 HSGP fusion-related guidance, and the following information provides a strategic outline that should be utilized to support this goal. In order to achieve a baseline level of capability, each fusion center must possess and/or prioritize efforts and expenditures to achieve the following capabilities:

Management/governance
- Defined management structure that governs intelligence activities
- Identification of core (permanent) and ad hoc stakeholders, including engagement with appropriate federal entities (Joint Terrorism Task Forces [JTTFs], Field Intelligence Groups [FIGs], High Intensity Drug

---

* US Department of Homeland Security, FY 2007 Homeland Security Grant Program, Supplemental Resource: Fusion Capability Planning Tool, January 2007.

Trafficking Area offices [HIDTAs], DHS field components such as Customs and Border Protection [CBP] and Immigration and Customs Enforcement [ICE], etc.)
- Governance structure (both multidisciplinary and multilevel of government) and/or some type of multidisciplinary advisory committee
- Defined goals and objectives to guide collection, analysis, dissemination, and use of intelligence
- Memorandums of understanding with all stakeholders
- Processes to add or remove stakeholders/partners as the fusion center matures
- Formal processes to define information and intelligence collection requirements

Fusion capability planning tool
- Defined mission statement
- Defined, agreed upon, and auditable privacy policy that covers collection, analysis, and dissemination
- Funding strategy that will cover operational costs for the next four years

Planning and requirements development
- Completion of a regional (or statewide) risk assessment (threat, vulnerability, and consequence)
- Capacity to identify patterns and trends reflective of emerging threats
- Collection requirements based on results of risk assessments
- Capacity to identify the circumstances or events (crime, public health, etc.) that represent indicators and/or precursors of threats
- Identification of the sources and/or repositories of data and information regarding indicators and precursors
- Collection of key information from existing sources
- Conducting public education and other activities necessary to enhance situational awareness by the public
- Training of frontline law enforcement and other public safety liaison personnel, including nontraditional partners and the public, so that they can better identify suspicious activities that may represent planning and/or operational activity by a terrorist group
- Mechanism to support reporting of collected information (9-1-1, tips line, fusion centers, Internet, etc.)
- Identification of regulatory, statutory, privacy, and/or other issues that impede collection and sharing of information, including an oversight mechanism to ensure individual privacy and civil liberties are protected
- Identification of and coordination with appropriate response and recovery personnel and operations centers to ensure fusion center activities are coordinated with and can be leveraged to support emergency operation activities, as appropriate, in the event of an incident

- Processes to obtain detailed knowledge from and provide information to homeland security partners and the private sector, including:
  - Vulnerabilities to possible terrorist attack
  - Assessment of the likelihood of attack
  - Likely methods of attack
  - Equipment and substances to carry out such an attack
  - Identification of planning activities

Collection
- Procedures and protocols to facilitate the communication of collection requirements to relevant local, state, and private sector entities
- Implementation of "situational awareness" activities (training, public education, etc.)
- Mitigating of impediments to collection
- Compilation of classified and unclassified data, information, and intelligence generated by individuals and organizations
- Integration of homeland security information (all-threats, all-hazards, and all-crimes) collection efforts with other reporting systems (9-1-1, 3-1-1, etc.)
- Establishment of process to identify and track reports of suspicious circumstances (e.g., preoperational surveillance, acquisition of items used in an attack)

Analysis
- Blending of data, information, and intelligence received from multiple sources
- Reconciliation, deconfliction, and validation of the credibility of data, information, and intelligence received from collection sources
- Evaluation and analysis of data and information using subject matter experts
- Identification and prioritization of the risks faced by the jurisdiction (state, locality, etc.)
- Production of value-added intelligence products that can support the development of performance-driven, risk-based prevention, protection, response, and consequence management programs
- Identification of specific preventive and protective measures to identify and disrupt potential terrorist attacks during the planning and early operational stages
- Appropriate training and certification of analysts

Dissemination, tasking, and archiving
- Identification of those entities and individuals (officials, executives, etc.) responsible for developing and implementing prevention, protection, response, and consequence management (public and private) efforts
- Provision of relevant and actionable intelligence in a timely manner to those entities responsible for implementing prevention, protection,

response, and consequence management efforts (public and private sector)
- Archiving of all data, information, and intelligence to support future efforts
- Access to Law Enforcing Online (LEO), the Homeland Security Information Network (HSIN), and other key information sharing systems
- Development of performance-based prevention, protection, response, and consequence management measures
- Capacity to track performance metrics associated with prevention, protection, response, and consequence management efforts
- Provision of feedback to collectors of information
- Access to a secret enclave where classified systems and secure rooms are available for use

Reevaluation
- Tracking of achievement of prevention, protection, response, and consequence management program performance metrics so as to evaluate impact on risk environment
- Update of threat, vulnerability, and consequence assessments so as to update risk environment

Modification of requirements
- Modification of collection requirements as necessary
- Communication of modifications to key stakeholders in a timely manner

The Fusion Center Guidelines define a fusion center as "a collaborative effort of two or more agencies that provide resources, expertise, and information to the center with the goal of maximizing their ability to detect, prevent, investigate, and respond to criminal and terrorist activity."* Some apply this definition broadly to include any multijurisdictional, anticrime, or response effort that may utilize intelligence and information, to include federally owned and operated collaborative efforts, like FBI-led Joint Terrorism Task Forces (JTTFs) or High Intensity Drug Trafficking Area (HIDTA).

Numerous national strategies have assessed the primary threat to U.S. national security as terrorism, at home and abroad. The National Strategy on Homeland Security provides that "the American People and way of life are the primary targets of our [terrorist] enemy and our highest protection priority." The means for combating this threat are broad and encompass all elements of national power, including nontraditional sectors.

From a law enforcement perspective, it has been argued that state and regional intelligence fusion centers, particularly when networked together nationally,

---

* US Department of Justice, Fusion Center Guidelines: Developing and Sharing Information and Intelligence in a New Era, August 2006, http://it.ojp.gov/documents/fusion_center_guidelines_law_enforcement.pdf.

represent a proactive tool to be used to fight a global adversary that has both centralized and decentralized elements. This network of fusion centers is envisioned as a central node in sharing terrorism, homeland security, and law enforcement information with state, local, regional, and tribal law enforcement and security officials. Today, there are over forty intelligence fusion centers across the country.*

These fusion centers are relevant to public safety EOC and private sector ECC operations because they represent a new resource of data, information, and an improved common operating picture of the critical infrastructure within their respective region. Further, many of these fusion centers have taken aggressive steps to reach out to business and industry partners to share sensitive information that contributes to more effective prevention and protection measures.

Fusion Liaison Officers (FLOs)

According to the Congressional Research Service (CRS, 2007) the relationship and role of the private sector is a function that most state fusion centers have yet to fully define and embrace. A number of the fusion centers surveyed have undertaken informational and security-related discussions with some of the major critical infrastructure owners and operators and data providers within their respective jurisdictions. However, while acknowledging that a comprehensive understanding of the risks to the state or region is impossible to attain without a viable relationship and consistent information flow between the center and the private sector entities within the center's jurisdiction, the vast majority of centers have yet to put the processes in place to support such an endeavor.†

FLO programs that foster the dynamic flow of suspicious activity reports to the counterterrorism/fusion center operational level are now being implemented. These initiatives provide near instantaneous early warning and situational awareness, facilitating the execution of informed preventative decisions and actions.

FLOs serve as the extra set of eyes in the community. Participants are trained to properly detect surveillance and suspicious activity, and to report such activity via the online suspicious activity reporting system, as well as through the established Homeland Security Hotline and/or other conventional/traditional means (e.g., calls to public safety 911 centers).

Fusion center analysts serve as contact points for the private sector organizations in reporting suspicious activity in and around their respective properties, serving to analyze, deconflict, and further disseminate all pertinent critical infrastructure intelligence data of significance to local and national homeland security. In addition, analysts provide regular sector-specific intelligence alerts and reports in the

---

* Congressional Research Service (CRS), Fusion Centers: Issues and Options for Congress, Washington, DC, July 6, 2007.
† Congressional Research Service, Fusion Centers: Issues and Options for Congress, Washington, DC, July 2007.

form of bulletins and briefings to provide continuous situational awareness to all private sector partners.

Public–Private Partnerships

The Infrastructure Security Partnership (TISP) was established following the tragic events of September 11, 2001, as a national forum for public and private sector organizations to collaborate on issues regarding the resilience of the nation's critical infrastructure against the adverse impacts of natural and manmade disasters. TISP members—who represent the design, construction, operation, and maintenance communities; local, state, and federal agencies; academe; and other organizations concerned about disaster preparedness, response, and recovery—work together to identify and develop cost-effective solutions by leveraging their collective resources, experience, technical expertise, research and development capabilities, and knowledge of public policy regarding natural and manmade disasters. Since its establishment, membership has grown to more than 100 organizations representing more than 1.5 million individuals and firms.

TISP has published *Regional Disaster Resilience: A Guide for Developing an Action Plan*, which is intended to provide a flexible, dynamic, high-level framework for use by all levels of government, service providers, and other organizations to create an action plan to help develop or improve comprehensive regional preparedness. The guide is intended to be as comprehensive as possible in offering the range of actions that organizations can collectively and individually take—on the basis of their perceived needs—to achieve regional disaster resilience. The focus is on bringing together public and private sector organizations as well as other entities with roles and missions or vested interests in disaster preparedness to address multihazards. The goal is to sensibly and cost-effectively secure the interdependent cyber and physical critical infrastructures.

Based on TISP regional resilience guidance an action plan is based on the following assumptions:

1. Environmental protection is integral to comprehensive regional preparedness and is key to ensuring that communities, states, and the nation can expeditiously respond to and recover from disasters of all types, particularly extreme events.
2. National and global infrastructures are increasingly complex and interconnected, resulting in physical and cyber vulnerabilities that we are only just beginning to understand. Stakeholder organizations are becoming increasingly aware of infrastructure interdependencies but need to broaden their knowledge of the extent of these linkages and their effects on responsibilities, operations, and business practices, particularly regarding large-scale and long-term disruptions.

3. An integrated and complementary cyber and physical approach is required to determine how best to secure interdependent infrastructures, ensure expeditious response and recovery, and build resiliency to address regional disasters. Consequently, there needs to be increased interaction among physical and cyber security personnel and emergency managers and operators to raise awareness of threats and vulnerabilities.

4. Today's preparedness needs require a comprehensive, multihazards regional approach that addresses natural disasters of all types, human error, systems failures, pandemics, and malevolent acts, including those involving cyber systems and weapons of mass destruction (chemical, biological, radiological, and nuclear devices).

5. Hurricane Katrina clearly demonstrated that existing federal, regional, state, and local disaster management plans need improvement to successfully deal with extreme disasters, natural or manmade. New thinking, approaches, training, and exercises as well as unprecedented intergovernmental collaboration and planning are required. This all must be accomplished in cooperation with the private sector and other key stakeholders.

6. The creation of regional public–private partnerships is necessary to bring together key stakeholders to build trust, foster information sharing and coordination, identify and assess vulnerabilities and other preparedness needs, and develop and implement solutions. Such partnerships should include all levels of government, utilities and other service providers, commercial enterprises (including businesses essential to localities; manufacturers; producers; processors; and distributors of important commodities and products), nonprofits, and community and academic institutions.

7. Extensive work has already been accomplished by the multihazards community that can be used to assist in preparedness for terrorist attacks.

8. Advances in information technology, engineering, materials, and biosciences as well as other disciplines are creating new vulnerabilities that we must anticipate.

9. Because of infrastructure interdependencies, protection of critical assets by means of physical security is only one important element of the holistic approach necessary to ensure essential services and products. There is also a need for cost-effective mitigation of potential and actual damage from disruptions, particularly those causing cascading effects that can incapacitate a region and impede rapid response and recovery.

10. Security and disaster resilience should be incorporated into cyber and physical systems in the development phase on the basis of assessed risk under various scenarios. Resilience can include system hardening, building in redundancies, implementing backup systems, and other mitigation measures.

11. Determining the criticality of infrastructure assets presents a major challenge, particularly for states and local governments that must make investment trade-offs on security and mitigation measures. Criticality is often in the eyes of the beholder and is dependent upon a given situation. The large and diverse

number of critical assets within a region, constrained state and local resources, and our need to gain better understanding of infrastructure interdependencies require the development of criteria for and a risk-based approach to identifying critical assets, infrastructure protection, and resiliency.

12. Environmental protection is integral to infrastructure protection and resiliency. Waste products and toxic holding sites, for example, should be considered security risks as well as environmental risks.

13. A major challenge is obtaining the necessary data on infrastructure interdependencies to enable the development of assessment and decision tools to provide greater understanding of associated cyber and physical vulnerabilities and how best to minimize them. Surmounting this challenge requires cooperation and finding ways to identify, collect, securely store, and share information provided by stakeholders that play significant roles in regional disaster preparedness.

14. Development and maintenance of mutual assistance agreements, user agreements, memorandums of understanding (MOUs), and other types of cooperative arrangements are essential to sound preparedness planning and disaster management. Such mechanisms enable jurisdictions, private sector organizations, and other stakeholders to resolve resource requirements and allocations, security and legal issues, sharing of proprietary information, and cost reimbursement in advance of emergencies.

15. Disaster response, recovery, and restoration are local and state missions, not the sole responsibility of the federal government.

16. Sorting out and defining roles and responsibilities—including determining who is in charge of particular functions—is fundamental to ensuring effective disaster preparedness, response, recovery, and restoration.

17. The federal government—historically through FEMA's regional offices and other regional federal offices—provides necessary guidance and assistance in preparedness and disaster response at the local and state levels. This regional approach should be continued and enhanced.

18. Integration of federal defense assets into regional preparedness in an appropriate manner is essential in addressing extreme disasters that require resources above and beyond those available at the state/provincial and local level.

19. Assuring supply chains and the delivery of critical products, materials, and components is essential to disaster resilience and the vitality of the industrial base, and has a direct and profound impact on regional and national economies and national security.

20. Where useful, codes, standards, and guidelines should be applied within and across organizations and jurisdictions to enhance security and preparedness, and to minimize costs.

21. Channels of communication must be established conscientiously and comprehensively to include representatives and spokespersons from all key stakeholders; must be tested frequently to improve and correct shortcomings and

to ensure that they work and have redundancy and resiliency to withstand infrastructure deterioration or destruction; must be maintained efficiently and regularly to ensure availability when needed; and must be owned by a well-established and defined entity with responsibility for administration and ongoing operations.

22. Law enforcement and public safety personnel at all levels should recognize the value of coordination with the public and private sectors when developing strategies to protect critical infrastructures. Proactive involvement with operators of critical infrastructure should provide the basis for sharing information that mitigates vulnerabilities and creates an awareness of threats to a system or facility.

23. Progress has been made to better manage large-scale events by means of the creation of NIMS and other measures. However, sorting out roles and responsibilities during major disasters and terrorist attacks remains one of the greatest challenges. Disasters know no jurisdictional boundaries, and key stakeholders must collectively define their responsibilities under various scenarios, taking into account the evolving roles of federal agencies—for example, DHS and the Department of Defense (DoD).

24. The private sector and nonprofit community possess a wealth of available resources and capabilities that must be incorporated into regional disaster response and restoration planning and activities.

25. Health care and public health organizations play a unique and highly important role in disaster response and should be integrated into disaster planning.

26. Community institutions, the general public, and individuals with special needs must be involved in planning and exercises, and particular emphasis should be placed on education and awareness of threats, impacts, and local emergency response procedures. Evacuation plans must be realistic in taking into account infrastructure interdependencies and individuals with special needs.

27. The media play a unique and integral role in disaster management, performing an information dissemination and education function—occasionally as first responders—and as essential stakeholders with business continuity needs. The media need to participate in preparedness planning and exercises.

28. Any system or technology solution requiring confidential or sensitive information must be developed from inception through production with the highest appropriate level of information security. At the same time, the need for information security must be balanced with stakeholder need to know to increase understanding of interdependencies and facilitate preparedness planning.

29. Costs for technology solutions, maintenance, and upgrades must be affordable to states, localities, and private sector organizations.

What national legislation has been shaped to support the public–private partnership for preparedness?

The United States Congress emphasized the importance of private sector preparedness and NFPA 1600 within The National Intelligence Reform Act of 2004. Within Section 7305, Private Sector Preparedness, Congress asks the Secretary of Homeland Security:

> To promote, where appropriate, the adoption of voluntary national preparedness standards such as the private sector preparedness standard developed by the American National Standards Institute and based on the National Fire Protection Association 1600 Standard on Disaster/Emergency Management and Business Continuity Programs.

Are external disasters that could affect a neighborhood, city, community, region, state, or nation considered in an enterprise's BCP? The grand challenge may simply be how many questions will be asked after an event concerning your BCP, response, mitigation and recovery if the business is on the precipice of collapse or someone has died. How many questions could be asked by a body of public inquiry or a 9/11 style national commission? Are you prepared to answer?

The Federal Emergency Management Agency (FEMA) has endorsed the ANSI "Organizational Resilience: Security, Preparedness, and Continuity Management Systems" (ASIS SPC.1-2009) as a referenced resource in its rollout of the Private Sector Preparedness (PS-PREP) program in 2010.

Summary

In this chapter we examined the important role that business continuity management and private sector emergency preparedness play in reducing the impacts of natural, technological, and human-induced disasters on business and industry and the communities they serve. The critical infrastructure sectors, national and sector-specific plans, and the ten professional practices of business continuity management were reviewed to understand the planning framework in which private sector emergency planners can operate to effectively design, train, and exercise to comprehensive response plans.

The role of the private sector in contemporary emergency management was reviewed with emphasis on the binding management protocols that must link private sector infrastructure owners and operators to the public safety community, including the National Incident Management System. The mission, roles, responsibilities, equipment and technologies supporting various types of private sector command centers were identified.

Emergent regional capabilities that provide tools, data, and operational assets for the private sector, including the DHS-sponsored fusion centers, were described in terms of information-sharing benefits. Evolving guidance for public–private partnerships for regional resilience was provided.

Taken together, the information in this chapter has been structured to enable you to understand and more carefully evaluate opportunities for integration of private sector preparedness programs with local, state, and federal public safety preparedness activities.

Appendix A: Organizing for Homeland Security and Emergency Management

Stephen Krill

Contents

We often chuckle at the question, "which came first, the chicken or the egg?" But in emergency management no one chuckles when it comes to its relationship—equal, superior or inferior—to homeland security. How we organize to prepare for and respond to all hazards is critical to our ability to save lives and protect property. Similarly, understanding the history and evolution of our organizational approaches to emergency management is important to enhancing national preparedness and optimizing the delivery of resources to people in need.

Why Homeland Security and Emergency Management?

In July 2002, President Bush released the National Strategy for Homeland Security (or "Strategy") and defined *homeland security* as "a concerted national effort to

363

prevent terrorist attacks within the United States, reduce America's vulnerability to terrorism, and minimize the damage and recover from attacks that do occur" (White House 2002, 2) Acknowledging the shared responsibility of homeland security, the strategy also emphasized the need for "compatible, mutually supporting state, local, and private-sector strategies." This shared responsibility is shown in the relationships being built across the nation as homeland security grows and evolves.

While many definitions exist, the National Response Framework defines *emergency management* as "the coordination and integration of all activities necessary to build, sustain, and improve the capability to prepare for, protect against, respond to, recover from, or mitigate against threatened or actual natural disasters, acts of terrorism, or other manmade disasters." (Department of Homeland Security n.d., 2) Throughout its history, the United States has faced many disasters, from natural disasters to hazardous materials releases to terrorist attacks. While the field of emergency management is more mature and acculturated than that of homeland security, much of the political focus has been directed toward securing the homeland since the terrorist attacks of 9/11.

Organizationally we find many differences between how the federal government structures its homeland security and emergency management resources over that by state and local governments. Given the importance of both disciplines, are there efficiencies to be gained with parallel organizations with homogeneous structures? Or is heterogeneity (read "autonomy") the optimal structure for securing the homeland?

Organizing for All-Hazard Emergencies

Dr. George Haddow notes that unlike other more structured disciplines, emergency management expands and contracts in response to events, legislative desires, and leadership styles (Haddow and Bullock 2003, 2). Such changes are present through both federal and state and local emergency management agencies. These changes suggest an inherent need for organizational adaptability, especially to correct problems identified during response or recovery operations.

Emphasizing the value of organizational evolution, in a 2006 study of federal emergency management and homeland security organizations, the Congressional Research Service concluded, "The present organization of federal emergency management functions is the latest development in a more than 50-year effort to find the most economical, efficient, and effective arrangements for protecting the nation from, and responding to, disasters" (Hogue and Bea 2006, 2). While the most economical, efficient, or effective arrangements are necessarily drivers behind changes in homeland security and emergency management structures, ensuring that lessons learned from past mistakes do not reoccur seems a stronger driver.

Especially at the federal level, changes in disaster response occurred following major disasters that caused significant consequences. Over the decades with a movement from military to civilian authorities for disaster response, and oftentimes a blending of the two, we find both centralized and decentralized organizational structures. Such structures also offer tension as civilians tend to resist military over-lays and soldiers tend to question civilian applications.

In their 2007 article, *When There Is No Cavalry*, which examined the need for government, industry, and civilian collaboration in emergency management, Himberger et al. found the command and control management structure in the federal government often embeds agencies with shifting, overlapping, and some-times ambiguous responsibilities with respect to disaster response. Referencing the Federal Emergency Management Agency (FEMA), the agency formed with a focus on civil defense and the Soviet nuclear threat. Following the end of the Cold War, FEMA emphasized disaster relief, recovery, and mitigation programs. After 9/11, with FEMA being absorbed into the Department of Homeland Security (DHS), questions arose around the similarities and differences between homeland security and emergency management. The authors noted in the kind of ambiguous struc-ture facing FEMA, the higher the degree of control exercised by any one agency or organization, the greater the potential for interorganizational conflict (Himberger et al. 2007, 8).

According to a 2001 study published by the Public Entity Risk Institute, an effective emergency management organization is defined to be one that "when applied during a disaster, will provide the levels of protection for life and property, and recovery assistance, which are acceptable to the citizens of the community" (Public Entity Risk Institute 2001, 3).

Returning to the National Strategy for Homeland Security, our overlapping federal, state, and local governance often challenges homeland security and emer-gency management. To address these challenges, the strategy emphasized the need to develop interconnected and complementary systems that reinforce rather than duplicate essential requirements, such as those needed for acts of terrorism versus natural or other types of disasters. To that end, the strategy advocated for effective preparation for catastrophic natural disasters and manmade disasters. Even though neither type of event has a terrorism nexus, preparing for such emergencies will nevertheless increase the security of the homeland.

Federal Organizational Structures

Federal Emergency Management Agency (FEMA)

Before FEMA's establishment in 1979, more than one hundred federal agencies were involved in some aspect of disasters, hazards, and emergencies. Fragmentation at the federal level led to multiple—sometimes duplicative—policies and programs around federal disaster relief. Seeking to simplify this complexity, the National Governor's Association asked President Carter to centralize federal emergency func-tions. According to Dr. Haddow, President Carter's Reorganization Plan Number

3 consolidated emergency preparedness, mitigation, and response activities into FEMA and made the FEMA Director a direct report to the President. (Haddow and Bullock 2003, 6). Further, FEMA focused on all-hazard emergencies. However, starting a new federal agency and managing sweeping organizational change did not come easy. Compounding its formation, responses to multiple disasters, such as the Loma Prieta Earthquake in 1989 and Hurricane Andrew in 1992, highlighted the need for additional organizational improvements at FEMA.

In 1993, President Clinton nominated James L. Witt as FEMA director, who finally brought experience as a state emergency manager to the position. Witt reformed disaster relief and recovery operations and emphasized preparedness and mitigation. He also reemphasized all-hazard emergency management. By most measures, Witt enjoyed a successful tenure at FEMA and strong relationships with his state and local counterparts. Eight years later, President Bush appointed Joe Allbaugh as FEMA director, and under his watch, the agency responded to 9/11. Even though FEMA had the role, following the Oklahoma City bombing and the embassy bombings in Nairobi and Kenya, as lead for consequence management to terrorist incidents, 9/11 tested the agency in unprecedented ways. Afterward, FEMA needed to focus more strongly on domestic terrorism than in its traditional role for all-hazard emergencies.

Department of Homeland Security (DHS)

Before DHS's establishment in 2003, homeland security activities were spread across more than forty federal agencies (Department of Homeland Security n.d.). Shortly after 9/11, President Bush issued Executive Order 13228, establishing the Office of Homeland Security within the Executive Office of the President and the Homeland Security Council. Both organizations advised the president on homeland security matters.

As previously mentioned, the White House in 2002 released the first National Strategy for Homeland Security. The strategy provided direction to federal government departments and agencies for homeland security and recommended steps that state and local governments, private companies and organizations, and individual Americans could take in securing the homeland.

Title I of the Homeland Security Act of 2002 established DHS with the mission of (1) preventing terrorist attacks within the United States; (2) reducing the vulnerability of the United States to terrorism; (3) minimizing the damage, and assisting in the recovery, from terrorist attacks that do occur within the United States; and (4) carrying out all functions of entities transferred to the department, including by acting as a focal point regarding natural and manmade crises and emergency planning. One of the first published DHS organizational charts showed five directorates: Border and Transportation Security, Emergency Preparedness and Response, Information Analysis and Infrastructure Protection, Management, and Science and Technology. While the act moved FEMA into DHS, with a direct line of report between the administrator and the secretary, it also placed the agency into the Emergency Preparedness and Response Directorate. This directorate consolidated all of DHS's preparedness programs, including grants, first-responder training, citizen awareness, public health, infrastructure, and cyber security.

Two years later, Secretary Michael Chertoff issued a revised DHS organizational chart that abolished the directorates for Border and Transportation Security, Information Analysis and Infrastructure Protection, and Emergency Response and Preparedness. With this change FEMA became a stand-alone agency within DHS, reporting directly to the Secretary of Homeland Security.

According to the Congressional Research Service, Chertoff's reorganization was not universally accepted. The National Emergency Management Association (NEMA), an association of state emergency directors, strongly criticized the reorganization, telling Congress that separating disaster planning from response and would "result in a disjointed response and adversely impact the effectiveness of departmental operations" (Hogue and Bea 2006, 21).

After the tragedy of Hurricane Katrina, the Post-Katrina Emergency Reform Act (Public Law 109-295) established new leadership positions within the department and gave additional responsibilities to FEMA. Specifically, the act transferred to FEMA all functions of the Preparedness Directorate, including the Office of Grants and Training, the United States Fire Administration, and the Office of National Capital Region Coordination. Further, the act made FEMA a distinct entity in the department and subject to reorganization only by statute.

State and Local Organizational Structures

Any discussion of state and local emergency response organizations must acknowledge federalism and states' rights. The Tenth Amendment decrees that "powers not delegated to the United States by the Constitution, nor prohibited by it to the States, are reserved to the States respectively, or to the people." Writing for the Heritage Foundation, James Jay Carafano and Richard Weitz (2006, 2) note America's system for emergency management reflects these principles, with each state deciding for itself the precise delineation of authorities and responsibilities for emergency response.

According to Buchalter and Miller (2007, 1), both state homeland security and emergency management directors share a common need to prevent and respond to acts of terrorism and other critical hazards. Searching for key words used in state and territorial statutes and agency mission statements, he found universal terms, such as preparedness and training; prevention; prompt, effective emergency response and recovery; minimization of injury; and identification of areas vulnerable to disaster and emergency.

Still, unlike their federal counterpart, state and local governments used differing organizational structures for homeland security and emergency management. In its 2008 State Homeland Security Directors Survey, the National Governors

Association (NGA) found that during the past five years, states have adjusted their governance structures and priorities to meet changing threats and better align federal grant program requirements (Ware 2006, 1).

The NGA survey showed states manage homeland security operations through a variety of structures that meet their own unique challenges and priorities:

1. An independent cabinet department, 28 percent
2. An advisory group of a larger cabinet-level department that coordinates budgetary and strategic decisions, 25 percent
3. A division or segment of a larger cabinet-level department, 30 percent

In a more detailed look at organizational structures, NEMA shows the variety of state emergency management agencies (see Table A.1). According to NEMA, the emergency management agency structures evolve, as well as the governor's influence and homeland security and emergency management organization.

Texas Division of Emergency Management

Looking more closely at a single state—Texas—shows this evolution in emergency management structure. The Texas Civil Protection Act of 1951 established the Division of Defense and Disaster Relief in the Governor's Office to handle civil defense and disaster response programs. In 1963, the division was collocated with the Department of Public Safety, and renamed the Division of Disaster Emergency Services in 1973. After several more name changes, in 2005 the organization was designated an operating division of the Texas Department of Public Safety. Legislation passed during the 81st session of the Texas Legislature in 2009 formally changed the name of the organization to the Texas Division of Emergency Management. Table A.2 illustrates the three main entities and their responsibilities for homeland security and emergency management in Texas (Texas Division of Emergency Management, 1–2).

Conclusions

Writing in *Homeland Security Affairs*, Bellevita (2008, 7) observes not all jurisdictions face the same threats or risks: Florida has hurricanes; Montana has wildfires; Iowa has floods; Arkansas has tornadoes; the northwest has earthquakes; the Great Lake states have severe winter storms; and New York City, Washington DC, and other major urban areas have greater risk for terrorist attacks. He argues that homeland security means different things to different jurisdictions, and he especially notes this construct when dealing with federal homeland security grants. To quote: "Homeland security should be a locally directed effort to prevent and prepare for incidents most likely to threaten the safety and security of its citizens."

Table A.1 Where State Emergency Management Agencies Fall in Various States

Governor	Adjutant General/ Military Affairs	Homeland Security	Public Safety	State Police	Other
13	18	4	12	2	4
Alabama	Alaska	Delaware	Massachusetts	Michigan	Colorado
Arkansas	Arizona	Indiana	Minnesota	New Jersey	Hawaii
California	Idaho	District of Columbia	Missouri		New Mexico
Connecticut	Iowa	Guam	Nevada		West Virginia
Florida	Kansas		New Hampshire		
Georgia	Kentucky		North Carolina		
Illinois	Maine		Ohio		
Louisiana	Maryland		South Dakota		
Mississippi	Montana		Texas		
New York	Nebraska		Utah		

Continued

Table A.1 (*Continued*) Where State Emergency Management Agencies Fall in Various States

Governor	Adjutant General/ Military Affairs	Homeland Security	Public Safety	State Police	Other
Oklahoma	North Dakota		Vermont		
Pennsylvania	Oregon		Virginia		
Northern Mariana Islands	Rhode Island				
	South Carolina				
	Tennessee				
	Washington				
	Wisconsin				
	U.S. Virgin Islands				

Source: National Emergency Management Agency.

Table A.2   The Three Entities and Their Responsibilities for Homeland Security and Emergency Management in Texas

Entity	Responsibilities
Governor	• Directs homeland security in the state and develops a statewide homeland security strategy • Deals with dangers to the state and people presented by disasters and disruptions to the state and people caused by energy emergencies
Governor's Office of Homeland Security[a]	• Provides policy guidance for state homeland security programs • Coordinates development and monitors implementation of the state homeland security strategy • Coordinates state homeland security programs with local governments, regional organizations, and federal agencies
Governor's Division of Emergency Management	• Administers comprehensive all-hazard emergency management program for the state and assisting cities, counties and state agencies in implementing their own emergency management program • Supports development and implementation of the Governor's Homeland Security Strategy • Maintains the State of Texas Emergency Management Plan and other specialized state plans • Conducts an extensive emergency management-training program for local and state officials and emergency responders • Provides threat awareness and preparedness educational materials for the public • Administers pre- and postdisaster programs to eliminate or reduce the impact of known hazards • Coordinates mobilization and deployment of state resources to respond to major emergencies and disasters

Source: Texas Division of Emergency Management.

[a] The Director of the Office of Homeland Security also serves as the Director of the Governor's Division of Emergency Management.

State and local governments will often use homeland security grants to improve their all-hazard emergency management programs in order to address their greatest risks.

What the NGA and NEMA surveys show is that state and local government have adapted—and will continue to adapt—their homeland security and emergency management structures to suit their own needs and to help facilitate their relationships with the federal government. This latter objective is most notably seen through the federal grants for homeland security and being able to obtain the highest possible funding level to support or sustain all-hazard emergency management programs. In many cases, these grants cover dual-use applications, supporting both all-hazard and terrorism preparedness activities.

Bellevita acknowledges the reason for dual-use applications. Simply put, states and communities do not have the resources to focus attention solely on terrorism (Bellevita 2008, 3). As a national issue, homeland security must involve everyone. State and local agencies need to fully leverage relationships across all levels of government, the private sector, and the public and optimize collaboration. This approach also strongly encourages the need to speak the same language and to understand each other's culture.

According to Clovis (2006, 18) also writing in *Homeland Security Affairs,* only through collaboration at all levels of government will America achieve the best possible level of homeland security preparedness. He asserts anything less will lead to inefficiencies and, worse, unnecessary risk.

While Clovis (2006, 10) also notes both Congress and the administration (in homeland security, principally through DHS) can use grants as a means for affecting change and influencing organizational behaviors, this approach runs counter to federalism. In support, Carafano (2006, 8) writes, "As with many other homeland security missions, applying—rather than trying to circumvent—the principles of federalism usually produces the best results."

Whichever organizational structure is selected, integrating homeland security and emergency management yields greater efficiencies than as stand-alone entities. Stephen Flynn (2004, 60) notes in *America the Vulnerable: How Our Government Is Failing to Protect Us from Terrorism*, "because of our event-driven approach to addressing vulnerabilities, Americans have failed to appreciate that security works best when it is integrated into the normal course of business. When security becomes a reactive enterprise, pursued only after threats become manifest, the efforts end up being costly, ugly, and largely ineffective."

References

Bellavita, Christopher. Changing Homeland Security: What Is Homeland Security? *Homeland Security Affairs* 4, no. 2 (June 2008).

Buchalter, Alice R., and Patrick Miller. *A Guide to Directors of Homeland Security, Emergency Management, and Military Departments in the States and Territories of the United States.* Washington: Library of Congress, 2007.

Carafano, James Jay, and Richard Weitz. *Learning from Disaster: The Role of Federalism and the Importance of Grassroots Response.* Backgrounder. No. 1923. Washington: The Heritage Foundation, March 21, 2006.

Clovis, Samuel H., Jr. Federalism, Homeland Security and National Preparedness: A Case Study in the Development of Public Policy. *Homeland Security Affairs* 2, no. 3 (October 2006).

Department of Homeland Security. *Brief Documentary History of the Department of Homeland Security: 2001-2008.* Washington, DC (n.d.).

Federal Emergency Management Agency. FEMA History. http://www.fema.gov/about/history.shtm (accessed October 18, 2009).

Federal Emergency Management Agency. NRF Resource Center. http://www.fema.gov/emergency/nrf/glossary.htm#E (accessed October 18, 2009).

Flynn, Stephen. *America the Vulnerable: How Our Government Is Failing to Protect Us from Terrorism.* New York: HarperCollins, 2004.

Haddow, George D., and Jane A. Bullock. *Introduction to Emergency Management.* Burlington, MA: Butterworth-Heinemann, 2003.

Himberger, Douglas, David Sulek, and Stephen Krill, Jr. *When There Is No Cavalry* (The Megacommunity Reader). McLean, VA: Booz Allen Hamilton, June 2007.

Hogue, Henry B., and Keith Bea. *Federal Emergency Management and Homeland Security Organization: Historical Developments and Legislative Options* (RL33369). Washington DC: Congressional Research Service, 2006.

National Emergency Management Association. *State Emergency Management Organizations.* http://nemaweb.org/default.aspx?ID=2076 (accessed October 18, 2009).

Public Entity Risk Institute. *Characteristics of Effective Emergency Management Organizational Structures.* Fairfax, VA: Public Entity Risk Institute, 2001.

Texas Division of Emergency Management. *State & Local Emergency Organizations.* http://www.txdps.state.tx.us/dem/documents/statelocalhsemorgsV4.doc (accessed October 18, 2009).

Ware, Will. *2008 State Homeland Security Directors Survey* (Issue Brief). Washington, DC: NGA Center for Best Practices, March 6, 2009.

White House. *National Strategy for Homeland Security.* Washington, DC: Office of Homeland Security, 2002.

Appendix B: Managing Spontaneous Volunteers

Michael Fagel

One thing is certain. Following a terrorist or disaster incident, you can expect to be overwhelmed by volunteers who are anxious to help. Volunteers can be a huge help—or a second disaster—based on how they are handled. Become aware of your jurisdiction's policy and your state's liability laws when making a decision about whether and how to use volunteer resources. Based on the information you gain, you may choose to send volunteers home. More likely, you'll decide to send them to a central location for coordination and assignment (see Figure B.1).

It is recommended that the issue of volunteers be addressed before an incident occurs and that a central coordination point be established for the community. Local and state Citizen Corps councils are an effective way to organize volunteers before an incident and manage them during an incident. Citizen Corps involves several programs including Community Emergency Response Teams (CERTs), Volunteers in Police Service (VIPs), Neighborhood Watch, the Fire Corps, and Medical Reserve Corps. If your community does not have an active Citizen Corps council, you could coordinate with the state Voluntary Organizations Active in Disaster (VOAD) or the American Red Cross, both of which are Citizen Corps affiliates. Otherwise, follow these suggestions:

- Develop a plan to deal with spontaneous volunteers in advance of an incident
- Form a volunteer coordination team
- Encourage the formation of disaster coalitions
- Identify potential partners
- Establish a volunteer reception center
- Designate an entity to manage groups of unaffiliated volunteers
- Review your jurisdiction's insurance policy with regard to volunteers
- Develop a training and exercise program for potential volunteers

Figure B.1 **Volunteers clear thousands of used sand bags from Clay County, Minnesota, as hundreds descended on Moorhead, Minnesota, to help in the aftermath of Red River flooding in 2009. (Courtesy FEMA, Mike Moore.)**

- Develop mutual aid systems
- Develop standardized public education and media messages to use before, during, and after disaster events
- Stress the need to avoid "disaster within the disaster" with regard to the involvement of unaffiliated volunteers
- Establish relationships with ethnically diverse media outlets and community leaders to ensure messages are designed to reach all segments of the community

Potential volunteer partners may include:

- Universities
- Youth groups
- Schools
- Civic associations

- Neighborhood groups
- Faith-based organizations
- Corporations and businesses
- Special needs groups
- Voluntary agencies
- Senior programs
- National service programs
- Hospitals
- Professionals in volunteer coordination

Appendix C: Emergency Management and the Media

Randall Duncan

Contents

Understanding and working with the media is an important part of an overall emergency management system. This relationship—between the emergency manager and the media—is one that has more opportunity to excel or fail than almost any other.

Let's begin our examination of the relationship between emergency management and the media by defining what the media are. Traditionally, we think of the media as consisting of newspapers, radio, and television. Newspapers have been the media staple since Johannes Gutenberg invented the modern printing press in 1450. Radio and television entered the world of media much more recently, but changed the way media operated and functioned within society by bringing news on a timelier basis (live reports) and adding the elements of voices (radio) and moving pictures (television).

More recently, the development of the World Wide Web, social network sites, and blogs has led yet another revolution in the way media impacts our lives.

In order to more fully understand the elements of the relationship between emergency management and the media, it is necessary to understand the characteristics of the various types of media. Let's begin our examination with the traditional media forms of newspapers, radio, and television.

Newspapers

Arguably, the first newspaper in the United States was called *Publick Occurrences Both Forreign and Domestick* (National Humanities Center 2006), shown in Figure C.1. It was published on September 25, 1690, and edited by Benjamin Harris. It only printed one issue and was banned four days after publication by the Governor and Council of Massachusetts (Massachusetts Historical Society 2010). The only surviving copy of the newspaper is in the Public Record Office in London (Library of Congress 2009).

Traditionally modern newspapers have published on a daily or weekly basis, depending on the size of the reading audience. Circulation of newspapers varies greatly. The newspapers with the three largest weekday circulations in 2009 were the *Wall Street Journal* (circulation 2,024,269), *USA Today* (circulation 1,900,116), and the *New York Times* (circulation 927,851). The newspaper showing the smallest circulation was the *Medina (NY) Journal Register* (circulation 2,117; Audit Bureau of Circulation 2009).

Newspapers have traditionally been viewed as providing more in-depth coverage than either radio or television because of the amount of space available in which to write the story. Newspapers also provided some of the first coverage of events far removed from the place where they were published by the mechanism of the telegraph (see the next section for more details). This allowed remote correspondents to send a story from far away back to the newspaper home office and created a style of journalistic writing known as the "inverted pyramid." The inverted pyramid style of writing called for the correspondent to relay the most important facts first, followed by those of lesser importance in the body of the story (Scanlon 2008).

Based on this information then, and as shown in Table C.1, we can anticipate what print organizations want in the way of news.

Radio

It is not possible to talk about the history of radio without mentioning the wired telegraph system. The telegraph was made practical within the United States by Samuel Morse, who did his first public demonstration of the device in 1838 (Smithsonian Institution 2010). In 1843, Congress provided funding to install a telegraph between Baltimore, Maryland, and Washington, DC. The Whig Party held its nominating convention in Baltimore on May 1, 1844, and selected Henry Clay as its nominee.

PUBLICK
OCCURRENCES
Both *FORREIGN* and *DOMESTICK*.

Boston, Thursday *Sept.* 25th. 1690.

IT is designed, that the Countrey shall be furnished once a moneth (or if any Glut of Occurrences happen, oftener,) with an Account of such considerable things as have arrived unto our Notice.

In order hereunto, the Publisher will take what pains he can to obtain a Faithful Relation of all such things; and will particularly make himself beholden to such Persons in Boston whom he knows to have been for their own use the diligent Observers of such matters.

That which is herein proposed, is, First, That Memorable Occurrents of Divine Providence may not be neglected or forgotten, as they too often are. Secondly, That people every where may better understand the Circumstances of Publique Affairs, both abroad and at home; which may not only direct their Thoughts at all times, but at some times also to assist their Businesses and Negotiations.

Thirdly, That some thing may be done towards the Curing, or at least the Charming of that Spirit of Lying, which prevails amongst us, wherefore nothing shall be entered, but what we have reason to believe is true, repairing to the best fountains for our Information. And when there appears any material mistake in any thing that is collected, it shall be corrected in the next.

Moreover, the Publisher of these Occurrences is willing to engage, that whereas, there are many False Reports, maliciously made, and spread among us, if any well-minded person will be at the pains to trace any such false Report so far as to find out and Convict the First Raiser of it, he will in this Paper (unless just Advice be given to to the contrary,) expose the Name of such person, as A malicious Raiser of a false Report. It is suppos'd that none will dislike this Proposal, but such as intend to be guilty of so villanous a Crime.

THE Christianized *Indians* in some parts of *Plimouth*, have newly appointed a day of Thanksgiving to God for his Mercy in supplying their extream and pinching Necessities under their late want of Corn, & for His giving them now a prospect of a very Comfortable Harvest. Their Example may be worth Mentioning.

'Tis observed by the Husbandmen, that altho' the With-draw of so great a strength from them, as what is in the Forces lately gone for *Canada*, made them think it almost impossible for them to get well through the Affairs of their Husbandry at this time of the year, yet the Season has been so unusually favourable that they scarce find any want of the many hundreds of hands, that are gone from them; which is looked upon as a Merciful Providence.

While the barbarous *Indians* were lurking about *Chelmsford*, there were missing about the beginning of this moneth a couple of Children belonging to a man of that Town, one of them aged about eleven, the other aged about nine years, both of them supposed to be fallen into the hands of the *Indians*.

A very Tragical Accident happened at *Water-Town*, the beginning of this Month, an Old man, that was of somewhat a Silent and Morose Temper, but one that had long Enjoyed the reputation of a Sober and a pious Man, having newly buried his Wife, The Devil took advantage of the Melancholy which he thereupon fell into, his Wives discretion and industry had long been the support of his Family, and he seemed hurried with an impertinent fear that he should now come to want before he dyed, though he had very careful friends to look after him who kept a strict eye upon him, least he should do himself any harm. But one evening escaping from them into the Cow-house, they there quickly followed him. found him hanging by a Rope, which they had used to tye their Calves withal, he was dead with his feet near touching the Ground.

Epidemical Fevers and Agues grow very common, in some parts of the Country, whereof, tho' many dye not, yet they are sorely unfitted for their imployments; but in some parts a more malignant Fever seems to prevail in such sort that it usually goes thro' a Family where it comes, and proves Mortal unto many.

The Small-pox which has been raging in *Boston*, after a manner very Extraordinary is now very much abated. It is thought that far more have been sick of it then were visited with it, when it raged so much twelve years ago, nevertheless it has not been so Mortal. The number of them that have

Figure C.1 Publick Occurrences Both Forreign and Domestick.

Table C.1   Items and Details in Which Print Media Will Most Likely Be Interested

Item	Explanation
Details	Print media wants to paint a picture in the reader's mind with words
Questions	Will be more oriented toward details
	How many times did the truck roll over?
	How far away from the edge of the road did it come to rest?
	Were there flames? If so, how high?
Background information	History related to an event
	Has this ever happened before?
	History of individuals involved in the event
Deadline	Traditional: usually daily; current policy may be impacted by newspaper Web site

This was the first news item relayed by telegraph (Smithsonian Institution 2010). In 1901 Guglielmo Marconi began developing what would become broadcast radio. He sent the Morse code signal for the letter S from a wireless transmitter in Poldhu, Cornwall, England, to a wireless receiver in Newfoundland, Canada (Public Broadcasting System 1998b). A few years later, on Christmas Eve 1906, some wireless telegraph operators aboard ships heard the Christmas carol "Silent Night" and a voice reading bible verses interrupt the Morse code they normally heard (Public Broadcasting System 1998a). This marked the first radio broadcast.

From these humble beginnings, radio had an impact on the way we listened to news and found out about other events. We could sit in our living rooms and hear the voices of residents, dictators from overseas, and Hollywood stars endorsing commercial products. Unlike newspapers, radios could bring us the sounds and words of a news event as they happened.

Modern radio stations are separated into various interest groups called "formats." Some of the formats in today's radio broadcasting include news/talk stations, music stations, public radio, and non-English radio.

Television

The first authorized broadcast of a television in the United States occurred July 2, 1928, in Wheaton, Maryland, a suburb of Washington, DC, by C. F. Jenkins ("What Television" 1928). The heyday of television may have come on the evening of March 7, 1955, when one in two Americans watched Mary Martin's portrayal of

Table C.2 Items and Details in Which Radio Will Most Likely Be Interested

Item	Explanation
Details	Radio news utilizes short, concise information in the voice of the newsmaker
	Typically, they are 10 to 15 second "actualities" or "sound bites"
News/talk	More stories; a little more depth than other radio formats
Public radio	Uses "natural sound"; records background of event happening with open mike
Non-English stations	Help to reach those who speak a language other than English within the community
Music stations	May or may not carry news; if they do, it typically consists of only short news items
Deadline	Hourly, depending on schedule of newscasts

Peter Pan on live television (Bogart 1958, 1). Other significant events that marked the impact of television on the way Americans received news included the coverage of such live events as the Kennedy–Nixon debates and mankind's first step on the moon. The addition of live images to go with sound literally brought the world into our homes every night. There are various types of news broadcasts on local television stations. They may range from spot news or breaking news of particular activities currently in progress to the regularly scheduled news programs. In addition, some local television stations may air special investigative reports or programs. Typically, local television stations will have an affiliation with a network and will present a network-oritinated program of national and international news.

Social Network Sites and the World Wide Web

No examination of media would be complete without exploring the impact of the World Wide Web and social media on individuals as well as the traditional media of newspaper, radio, and television.

The World Wide Web, as we know it today, first became reality in 1990 with the release of a point-and-click hypertext editor called "World Wide Web" (Berners-Lee 1998). In the ensuing years, the number of Web sites, their functionality, and the pure amount of information have exploded. Of course, with such rapid expansion, there is a need for a buyer-beware approach by users. It is as easy to find an academically reputable and accurate source on the World Wide Web as it is the

Table C.3 Items and Details in Which Television Media Will Most Likely Be Interested

Item	Explanation
Details	Video of the event may determine whether or not there is a story
	A story that otherwise would not make the news may become a story if there is video
	Similarly, a story of real importance may not make the news if there is no video
Types of news	Local news includes spot news; regular local news; investigative reports; feature news programs (either local or network); national news; and international news
Deadline	Various depending on the type of news that will be broadcast; major deadlines are typically for the evening news broadcast and the late night news broadcast

lunatic ravings of fringe elements. It is up to the user to find and place the appropriate value on sources available through this medium.

Aside from the ease of accessing information on the World Wide Web, people soon found that it was becoming a tool for personal communication and networking between friends, leading to the development of social network sites. Social network sites are defined as

> web-based services that allow individuals to (1) construct a public or semi-public profile within a bounded system, (2) articulate a list of other users with whom they share a connection, and (3) view and traverse their list of connections and those made by others within the system. The nature and nomenclature of these connections may vary from site to site. (Boyd and Ellison 2007)

We also think of social network sites as social media—a way to convey information to friends and learn about information from friends and others' opinions. Sites like Twitter, Facebook, MySpace, and blogs have become a way to share information, opinions, and even reflect on news events.

These same sites have had an impact on the way traditional media—newspapers, radio, and television—interact with their viewers and listeners. As a result, newspapers now also shoot video of news stories and provide it to their readers through the mechanism of their Web site. Radio stations do the same thing. Television stations now write news stories and publish them, similar to newspapers, on their Web sites. This has had a major impact on the traditional deadlines for the various forms of media.

Yet another World Wide Web-based phenomenon has emerged recently—the weblog or blog. This phenomenon has blurred the distinction between traditional press and bloggers:

> In the media world, it used to be clear who was in the news business and who was not. News businesses provided news, non-news businesses did not. Reporters worked for companies who were in the business to provide news. News businesses got paid, usually by advertisers, to collect, package and distribute information of interest to news audiences. Non-news businesses or organizations exist for other purposes—perhaps to deliver public service such as environmental protection, or to produce commercial goods such as fertilizer. Providing news for these businesses is simply not their reason for existence. In an instant news world, that distinction is becoming increasingly fuzzy.
>
> One of the most significant trends to come out of the collection of technologies we call the Internet, is the emergence of citizen journalists. 'Blogging,' from the term 'Weblog,' which used to describe people who would record and publish what they discovered on the Internet, reflects the ease with which almost anyone who writes today can also publish. As mentioned earlier, some of the bloggers have accumulated audiences in the millions and have influence as great as any of the celebrity journalists that used to be staples of our early evening hours at home. (Baron 2006, 47)

Because of these factors, the emergency manager or spokesperson for the jurisdiction has to keep in mind that there are now more audiences for the information prepared for the traditional media. There are the families of those directly impacted by the event or emergency, those in the immediate area of the emergency or disaster, and the "traditional media," along with the citizen journalist.

Dealing with the Media in a Crisis

To begin our discussion about dealing with the media in a crisis, we need to understand some of the basics about communication. We first learn to communicate as babies, before we can even begin to say words. We communicate with gestures and sounds, through a thing called nonverbal behavioral clusters. We'll discuss this in more detail shortly.

Communications are extremely complicated for such a seemingly simple thing. The act of communications starts as an idea in our brain. That idea wishes to be expressed or communicated to someone. It must then make its way through the filters of our belief system and perceptions. Then it must be encoded (either in speech or writing) and then broadcast to a receiver (a reader or listener). The receiver has to get the message, decode it, and run the decoded material through his or her own filters of belief systems and perceptions in order to understand the idea we originally wished to communicate. An understanding of the complications associated with the process of how we communicate allows for a new appreciation of a statement as seemingly simple as, "Pass the salt, please."

Nonverbal behavioral clusters associated with how we say and express things are more important in conveying meaning than the words we actually say (Blatner 2009). Because these nonverbal behavioral clusters typically convey a larger percentage of our communication than the specific choice of words, we tend to place more faith in the way the message is expressed. When the message being conveyed to an audience by a speaker's words is in conflict with his or her nonverbal behavioral clusters, the audience will not believe the speaker. As a simple thought experiment, recall the last time you observed a person on television and your reaction to that person was the thought that you didn't believe a word he or she said. The odds are, you felt that way because there was a conflict between the words of the speaker and his or her nonverbal behavioral clusters.

The normal process of communications takes a slight detour under a crisis situation. When a crisis is in progress, we need to provide assistance to the elected official or spokesperson to make sure we don't allow circumstances to take away from the messages we need to communicate to the public typically through the media. In other words, we need to avoid media pitfalls. Table C.4 is adapted from unpublished material from Dr. Vincent Covello, founder and director of the Center for Risk Communiction, and is used with permission.

In a crisis situation, the media follow certain patterns. Those patterns include:

- Searching for background information on the incident
- Dispatching reporters to the scene
- Obtaining access to the scene or the official spokesperson
- Dramatizing the situation
- Expecting a briefing complete with written information
- Expecting you to panic
- Becoming confused by technical information
- Exhausting resources
- Sharing information among themselves
- Acting professional and expecting the same
- Providing filler for stories if credible information is not available

The Joint Information System (JIS)/ Joint Information Center (JIC)

As you have read previously, the National Incident Management System (NIMS) was developed at the direction of Homeland Security Presidential Decision Directive 5

Table C.4   How to Handle Media Situations in a Crisis

Do	*Don't*
Define all technical terms and acronyms (jargon)	Use language that may not be understood by even a portion of your audience
If you use humor, direct it at yourself	Use humor in relation to safety, health, or environmental issues
Refute negative allegations without repeating them	Refer to national problems, e.g., "This isn't Love Canal"
Use visuals to emphasize key points	Rely entirely on words
Remain calm, use the question or allegation as a springboard to say something positive	Let your feelings interfere with your ability to communicate positively
Ask whether you made yourself clear	Assume you have been understood
Use examples, stories, and analogies to establish a common understanding	Talk only in abstractions
Be sensitive to nonverbal messages you are communicating	Allow your body language or your position in the room to be inconsistent with your message
Make them consistent with what you are saying	Dress inconsistently with your message
Attack the issue	Attack the person or the organization
Promise only what you can deliver	Make promises you can't keep
Set, then follow strict deadlines	Fail to follow up on those promised items
Emphasize achievements made and ongoing efforts	Say there are no guarantees
Refer to the importance you attach to health, safety, and environmental issues; your moral obligation to public health outweighs financial considerations	Don't refer to the amount of money spent as a representation of your concern
Use personal pronouns (e.g., I, we)	Take on the identity of a large organization

Continued

Table C.4 (*Continued*)   How to Handle Media Situations in a Crisis

Do	Don't
Take responsibility for your share of the problem	Try to shift blame or responsibilities to others
Assume everything you say and do is part of the public record	Make side comments or "confidential" remarks
Discuss risks and benefits in separate communications	Discuss your costs along with risk levels
Use risk comparisons to help put risks in perspective	Compare unrelated risks
Stress that the true risk is between zero and the worst-case estimate	State absolutes or expect laypersons to understand
Emphasize performance, trends, and achievements	Mention or repeat large, negative numbers
Focus your remarks on empathy, competence, honesty, and dedication	Provide too much detail or get drawn into protracted technical debates
Keep presentation to fifteen minutes total	Ramble or fail to plan the time well
Keep answers to two minutes maximum	Tell people more than they want

(HSPD-5). The original NIMS document was developed in 2004, updated in 2006, and updated again in 2008 (FEMA 2008). Public information, and the process of establishing the system to collect, integrate and coordinate it, is defined as a part of NIMS Component IV–Command and Management (FEMA 2008, 70–74). The overall system is called the Joint Information System (JIS), and the specific place where this process happens is called the Joint Information Center (JIC).

Typically, this process involved a public information officer, who supports the incident command structure and who is also a member of the command staff. The responsibilities of this position typically include:

■ Responding to inquiries from the media, public, and elected officials
■ Supervising the process of collecting, integrating, and coordinating information for emergency public information
■ Supervising the process of collecting, integrating, and coordinating information for warning information
■ Monitoring for rumors and responding to them
■ Relations with the media

- Creating coordinated and consistent messages through
  - Identifying key information to be communicated to the public
  - Creating the message that provides the key information in a clear and easily understood method
  - Prioritizing messages so that the most important gets out first and that the public is not overwhelmed with the amount of information
  - Verifying the accuracy of information
  - Making sure the message gets out through the most effective means available

The overall JIS provides the means to coordinate the messages being released to the public by all elements of government involved in the disaster response—whether multiple local jurisdictions, or local, state, and federal governments; multiple disciplines involved in the response; nongovernmental organizations involved with the response; and the private sector. This coordination is particularly important because all the voices involved in the disaster should be providing substantially the same message.

One of the important elements to keep in mind when dealing with a large emergency response situation is that different disciplines may have their own spokesperson or PIO present, and different jurisdictions may have their own spokesperson or PIO present as well. It is possible that these PIOs may serve as the basis for the JIC staff. Remember that each of these officials will have the primary responsibility for making sure the story—as it relates to their agency, discipline, or jurisdiction—gets out to the media. But there is no reason they can't work together and collaborate to establish the JIS and provide the personnel for the JIC.

The JIC is basically an instrument to help facilitate the processes that take place within the JIS, much like the EOC is an instrument to help facilitate the processes that take place within the local emergency operations plan (LEOP). As a result, there are other parallels between these two elements of public information. The elements of the JIS must be worked out well in advance of the occurrence of a disaster. The system plans and processes, much like the roles and responsibilities within the LEOP, must be worked out and understood in advance of their application.

There should also be consideration given as to what kinds of triggers might initiate the activation of the JIC. Some suggestions might include:

- The creation of a standard operating procedure or guideline that defines the opening of the facility; can be modeled after the existing document for the activation of the emergency operations center
- An analysis of the potential impact of the incident
- An analysis of the potential media interest in the incident
- The potential duration or the response and recovery phases of the emergency or disaster

Table C.5   Key Points Regarding the JIS and JIC

Joint Information System (JIS)— Provides a Structure and System for:	*Joint Information Center (JIC)— Provides a Place to:*
Developing and delivering coordinated interagency messages	Centrally facilitate operation of the JIS during and after an incident
Creating, recommending, and executing public information plans and strategies	Increase information coordination
Advising incident commander about incident relevant public affairs issues	Reduce misinformation
Monitoring and correcting erroneous information circulating among the media or the public	Maximize resources for dealing with the public and the media
Be adaptable to the size and scale of the incident, from 3 PIOs at the scene to 150 PIOs at a major disaster from multiple locations	Provide "one-stop shopping" for the media

Other items noteworthy about the JIS and JIC are outlined in Table C.5.

Table C.6 presents a number of different types of JICs. It is adapted from information found in FEMA G290 Basic Public Information Officer Course (FEMA 2009). Figure C.2 provides a helpful JIC Readiness Assessment.

Last, a way for emergency managers to endear themselves to their local politician or public official is to provide them with a frank but informational "cheat sheet" on how to handle media in the event of a disaster (see Figure C.3).

Table C.6   Different Types of JICs

JIC Type	Description
Incident	Typically, an incident-specific JIC is established at a single, on-scene location in coordination with federal, state, tribal, and local agencies or at the national level, if the situation warrants. It provides easy media access, which is paramount to success. This is a typical JIC.
Virtual	A virtual JIC is established when a physical colocation is not feasible. It connects PIOs through e-mail, cell or landline phones, faxes, video teleconferencing, Web-based information systems, and so forth. For a pandemic incident where PIOs at different locations communicate and coordinate public information electronically, it may be appropriate to establish a virtual JIC.
Satellite	A satellite JIC is smaller in scale than other JICs. It is established primarily to support the incident JIC and to operate under its direction. These are subordinate JICs, which are typically located closer to the scene.
Area	An area JIC supports multiple-incident ICS structures that are spread over a wide geographic area. It is typically located near the largest media market and can be established on a local, state, or multistate basis. Multiple states experiencing storm damage may participate in an area JIC.
Support	A support JIC is established to supplement the efforts of several incident JICs in multiple states. It offers additional staff and resources outside of the disaster area.
National	A national JIC is established when an incident requires federal coordination and is expected to be of long duration (weeks or months) or when the incident affects a large area of the country. A national JIC is staffed by numerous federal departments and agencies, as well as state agencies and nongovernment organization (NGOs).

Plans		
Do you have systems and procedures for:	Yes	No
• Developing an emergency response or crisis communications plan for public information and media relations?	☐	☐
Does your emergency response or crisis communications plan have systems and procedures for:	Yes	No
• Designating and assigning line and staff responsibilities for the public information team?	☐	☐
• Identifying and updating current contact numbers for PIO staff and other public information partners in your plan?	☐	☐
• Identifying and updating current contact numbers for regional and local news media (including after-hours news desks)?	☐	☐
• Establishing the JIC at the emergency operations center (if activated)?	☐	☐
• Securing needed resources (space, equipment, people) to conduct the public information operation during an incident 24 hours a day, using such mechanisms as memorandums of understanding, contracts, etc.?	☐	☐
• Creating messages for the news media and the public under severe time constraints, including methods to clear these messages within the emergency response operations of your organization (including multijurisdiction and/or agency cross-clearance)?	☐	☐
• Disseminating information to news media, the public, and partners (e.g., Web site capability 24/7, listservs, broadcast fax, printed news releases, door-to-door leaflets, etc.)?	☐	☐
• Verifying and clearing/approving information prior to its release to the news media and the public?	☐	☐
• Operating a public inquiry hotline with trained staff available to answer questions from the public and control rumors?	☐	☐
• Activating the Emergency Alert System, including the use of prescripted messages?	☐	☐
• Coordinating your public information systems planning activities with other response organizations?	☐	☐
• Testing the plan through drills and exercises with other response team partners?	☐	☐
• Updating the plan as a result of lessons learned through drills, exercises and incidents?	☐	☐

Figure C.2   JIC Readiness Assessment. (From Federal Emergency Management Agency (FEMA), *G290 Basic Public Information Officer Training*, A39–A42, 2009.) *Continued*

People		
Do you have systems and procedures for:	Yes	No
• Identifying staffing capabilities needed to maintain public information operations for 24 hours per day for at least several days? (Note: Staff may include regular full- and part-time staff as well as PIOs from other agencies or departments, disaster employees, volunteers, etc.)	☐	☐
• Establishing and maintaining agreements for acquiring or borrowing temporary staff? (Note: Such agreements may be mutual aid arrangements or memorandums of understanding.)	☐	☐
• Granting emergency authority to hire or call up temporary staff or those on loan from other organizations?	☐	☐
• Establishing and maintaining job descriptions and qualifications for individuals serving as your organization's PIO and other roles during an incident?	☐	☐
• Assigning a staff member and at least one alternate the role and responsibilities of PIO?	☐	☐
• Determining if the assigned PIO(s) is qualified? Sample qualifications include: – Experience and skills in providing general and emergency public information. – Ability to represent your organization professionally (can articulate public information messages well when dealing with the media and the public, and can handle on-camera interviews). – Written and technical communication skills (writing/editing, photography, graphics, and Internet/Web design proficiency). – Management and supervision experience and skills needed to run a JIC.	☐	☐
• Establishing and maintaining a list of language translators available to assist with public information? (Note: Such network should include sign language interpreters and individuals capable of writing and speaking the non-English language(s) used by individuals in your jurisdiction.)	☐	☐
• Establishing and maintaining working relationships with PIO partners from other organizations that you might need to work with during an incident (e.g., PIOs from other jurisdictions, other government agencies or departments, nongovernmental organization, and private entities)?	☐	☐
• Developing and maintaining working relationships with your local and regional media, and established procedures for providing information to those media entities effectively and efficiently during incidents?	☐	☐

Figure C.2 (*Continued*) JIC Readiness Assessment. (From Federal Emergency Management Agency (FEMA), *G290 Basic Public Information Officer Training*, A39–A42, 2009.) *Continued*

Logistics		
Do you have a go-kit for PIO use during an incident, including:	Yes	No
• Laptop computer capable of linking to the Internet/e-mail?	☐	☐
• Cell or satellite phone, pager, and/or PDA/palm computer with wireless e-mail capability?	☐	☐
• Digital camera, photo storage media, and charger/backup batteries?	☐	☐
• Flash drives, CDs and/or disks containing the elements of the crisis communication plan (including news media contact lists, PIO contact lists, and information materials such as topic-specific fact sheets, backgrounders, talking points, and news release templates)? Remember: Redundancy is important in case the computer you are using doesn't have a USB port, CD, or floppy drive.	☐	☐
• Office supplies such as paper, pens, self-stick notes, etc.?	☐	☐
• Manuals and background information necessary to provide information to the media and the public (e.g., your Smart Book)? (Note: A Smart Book is a compilation of factual information assembled about your jurisdiction, such as population, number of schools and hospitals, size and description of geographic or infrastructure features, etc.)	☐	☐
• Hard copies of all critical information?	☐	☐
Do you have systems for:	Yes	No
• Acquiring and maintaining go-kits with a funding mechanism (e.g., credit card) that can be used to purchase operational resources? (Note: A go-kit is a mobile response kit that allows PIOs to maintain communications in the event that they are working outside of their normal place of operation.)	☐	☐
• Ensuring PIOs can access the go-kit when serving at an incident?	☐	☐
• Acquiring and maintaining portable communications equipment, critical up-to-date information, and supplies?	☐	☐
• Acquiring and maintaining essential media production equipment (cameras, digital storage, laptops, etc.)?	☐	☐
• Acquiring and maintaining a Smart Book (or equivalent technologies) to assist PIOs in accurately informing the media and the public during an incident?	☐	☐
• Identifying a dedicated location to house the JIC? (Note: The location selected must be wired for telephone, Internet access, cable, etc.)	☐	☐

Figure C.2 (*Continued*) JIC Readiness Assessment. (From Federal Emergency Management Agency (FEMA), *G290 Basic Public Information Officer Training*, A39–A42, 2009.) *Continued*

Logistics, cont.		
Do you have systems for:	Yes	No
• Securing and maintaining the necessary JIC equipment and supplies to allow information to be disseminated to the media and the public?	☐	☐
• Inventorying and restocking the PIO go-kit after an incident?	☐	☐
• Inventorying and restocking JIC equipment and supplies after an incident?	☐	☐
• Periodically updating your Smart Book with current information?	☐	☐
Do you have equipment and supplies needed for a JIC, including:	Yes	No
• Computers on a LAN with Internet access and e-mail listservs designated for news media and partner entities?	☐	☐
• Laptop computers?	☐	☐
• Electric and manual typewriter(s) in case of power outage or other problems that interfere with computer/printer usage?	☐	☐
• Fax machine preprogrammed for broadcasting fax releases to news media and partner entities?	☐	☐
• Printers and copy machines, with supplies such as toner and paper?	☐	☐
• Paper shredder and trash bags?	☐	☐
• Televisions with access to cable hookups and VHS VCRs or other recording media?	☐	☐
• Cell or satellite phones, pagers, and/or PDAs/palm computers with wireless e-mail capability?	☐	☐
• Digital camera, photo storage media, and charger/backup batteries?	☐	☐
• Audio recorder and batteries?	☐	☐
• Flash drives, CDs, and/or disks containing the elements of the crisis communication plan (including media contact lists, PIO contact lists, and information materials such as topic-specific fact sheets, backgrounders, talking points, and news release templates)?	☐	☐
• Office furniture/accessories such as desks, chairs, file cabinets, bulletin boards, white boards, trash cans, lights, in/out baskets, landline phones, clocks, large calendars, etc.?	☐	☐

Figure C.2 (*Continued*) JIC Readiness Assessment. (From Federal Emergency Management Agency (FEMA), *G290 Basic Public Information Officer Training,* A39–A42, 2009.) ***Continued***

Logistics, cont.		
Do you have equipment and supplies needed for a JIC, including:	Yes	No
• Audio equipment and furniture necessary for conducting news conferences (e.g., wireless microphones, lectern, multibox, etc.)?	☐	☐
• Office supplies (e.g., white and colored paper, pens, self-stick notes, folders, blank tapes, binders, overnight mail supplies, tape, poster board, erasable and permanent markers, chart paper, easels, staplers and staples, press kit folders, binders, computer disks/CDs, hole punch, organization logo on stickers, letterhead, postage stamps, etc.)?	☐	☐
• Manuals, directories, and background information necessary to provide information to the media and the public (e.g., your Smart Book)?	☐	☐
• Hard copies of all critical information?	☐	☐

Figure C.2 (*Continued*) **JIC Readiness Assessment. (From Federal Emergency Management Agency (FEMA),** *G290 Basic Public Information Officer Training,* **A39–A42, 2009.)**

Dear Chief Elected Official,

Here are the issues that you will need to know immediately in order to successfully deal with the media while in the midst of a disaster or emergency:

- Make sure you have command of your own emotions—don't panic in front of the media. However, it is appropriate to show concern and sympathy for those involved in the disaster or emergency.
- Make sure you have a complete understanding of what is known about the incident at this time. Concentrate on the basics—who, what, where, when, and why—before you speak with the media.
- If you don't know the information asked for in a question—don't lie!
- If you don't know the information asked for in a question—don't make information up!
- If you don't know the information asked for in a question—indicate that you will find out, and then actually make the effort to follow-up and get back in touch with the questioner.
- Make sure that the media present understand there will be a central location for the release of information (typically the Joint Information Center).
- Make sure that your staff supports the Joint Information System (JIS) process.
- Make sure you thoughtfully examine the situation to determine if you are the right person to be speaking on a topic. If you aren't, make sure that you and your staff find the right person to speak on the right issues.
- Never say, "No comment."
- If you can't answer a specific question, make sure the media know why you can't answer that question. For example the reason might be, "I'm sorry, but we simply don't have that information right now." Or, it might be, "The person who can answer your question is Police Chief Jones. Chief Jones, please step forward and provide what information you can on this issue."
- Treat every camera and recording device as if it is turned on the entire time it is in the same room with you.
- Everything you say will be "on the record."
- Make sure you have thought about the message you want to convey. Convey that message. When you're done conveying the message, then leave.
- Understand the different types of questions you will get from different media sources.
  - Typically, radio will want your voice, live, giving the details—but probably in a very short "sound bite."
  - Typically, television will want pictures or video of the incident.
  - Typically, newspapers will want more descriptive language and more details about the incident.
  - Don't forget that you will have new kinds of journalists in the crowd—"bloggers."
- First and foremost when dealing with the media—be honest!

I hope these briefing points will be of assistance.

Sincerely,
Your Emergency Manager

Figure C.3   Cheat sheet for public officials.

References

Audit Bureau of Circulation. 2009. US Newspapers, Search Results. http://abcas3.accessabc.com/ecirc/newstitlesearchus.asp (accessed April 15, 2010).

Baron, G. R. 2006. *Now Is Too Late2: Survival in an Era of Instant News*. Bellingham, WA: Edens Veil Media.

Berners-Lee, T. 1998, May 7. "The World Wide Web: A Very Short Personal History." World Wide Web Consortium (W3C). http://www.w3.org/People/Berners-Lee/ShortHistory (accessed April 23, 2010).

Blatner, A. (2009, June 29). "About Nonverbal Communications." http://www.blatner.com/adam/level2/nverb1.htm (accessed April 21, 2010).

Bogart, L. 1958. *The Age of Television*. New York: F. Unger.

Boyd, D. M. 2007. "Social Network Sites: Definition, History, and Scholarship." *Journal of Computer-Mediated Communication* 13(1), article 11, http://jcmc.indiana.edu/vol13/issue1/boyd.ellison.html (accessed April 21, 2010).

Federal Emergency Management Agency (FEMA). 2009. *G290 Basic Public Information Officer Training*. Washington, DC: Federal Emergency Management Agency.

Federal Emergency Management Agency (FEMA). 2008. "National Incident Management System." http://www.fema.gov/pdf/emergency/nims/NIMS_core.pdf (accessed April 23, 2010).

Library of Congress. (2009). "Eighteenth-Century American Newspapers in the Library of Congress." http://www.loc.gov/rr/news/18th/200.html (accessed April 15, 2010).

Massachusetts Historical Society. 2004. "Premier Issue of the Boston News-Letter." http://masshist.org/objects/2004.april.cfm (accessed April 15, 2010).

National Humanities Center. 2006. *Publick Occurrences Both Forreign and Domestick*. http://nationalhumanitiescenter.org/pds/amerbegin/power/text5/PublickOccurrences.pdf (accessed April 15, 2010).

Public Broadcasting System. 1998a. "KDKA Begins to Broadcast 1920." A Science Odyssey: People and Discoveries. http://www.pbs.org/wgbh/aso/databank/entries/dt20ra.html (accessed April 16, 2010).

Public Broadcasting System. 1998b. "Marconi Receives Radio Signal over Atlantic 1901." A Science Odyssey: People and Discoveries. http://www.pbs.org/wgbh/aso/databank/entries/dt01ma.html (accessed April 16, 2010).

Scanlon, C. (2008). "The Inverted Pyramid Structure." Purdue Online Writing Lab: http://owl.english.purdue.edu/owl/resource/735/04/ (accessed April 16, 2010).

Smithsonian Institution. (2010). "History Wired: A Few of Our Favorite Things." National Museum of American History, Smithsonian Institution: http://historywired.si.edu/detail.cfm?ID=324 (accessed April 16, 2010).

"What Television Offers You." 1928. *Popular Mechanics* 50(5): 820–824.

Index